T0295461

Understanding Nuclear Reactors

Understanding Nuclear Reactors

Global Warming and The Hydrogen Strategy

BRIAN HOOTON

OXFORD

UNIVERSITY PRESS

OXFORD
UNIVERSITY PRESS

Great Clarendon Street, Oxford, OX2 6DP,
United Kingdom

Oxford University Press is a department of the University of Oxford.
It furthers the University's objective of excellence in research, scholarship,
and education by publishing worldwide. Oxford is a registered trade mark of
Oxford University Press in the UK and in certain other countries

Published in the United States of America by Oxford University Press
198 Madison Avenue, New York, NY 10016, United States of America

British Library Cataloguing in Publication Data

Data available

Library of Congress Control Number: 2023946364

ISBN 9780198902652
ISBN 9780198902669 (pbk.)

DOI: 10.1093/oso/9780198902652.001.0001

Printed and bound by
CPI Group (UK) Ltd, Croydon, CR0 4YY

MIX
Paper | Supporting
responsible forestry
FSC
www.fsc.org   FSC® C013604

Preface

The first nuclear age came to an end following the disaster at Chernobyl. No politician would dare to put the word 'nuclear' on any agenda; it would only lose votes. The number of reactors starting to be built in the USA, the world leader in nuclear power, dropped to zero in 1994. As a nuclear physicist, I was sad to see the demise of the nuclear industry in the UK since we had led the way, in 1956, by building Calder Hall, the world's first nuclear reactor to send electricity into the national grid. I saw little prospect for any real new interest in nuclear power since the voting population were so against it, and it seemed to be so expensive when the enormous costs of waste management and decommissioning were taken into account. It would never be able to compete with fossil fuel as a source of energy.

The change came with the formation of the UN Framework Convention on Climate Change (UNFCCC), established in 1994, with a Treaty endorsed by 165 signatories. It anticipated that global warming would become a threat to our way of life. Many scientists were still sceptical about atmospheric CO_2 being the cause of global warming and were saying it was due to the ending of an ice age. Eventually, they started to look at the evidence and moved into the camp that said that global warming and climate change was indeed due to our extensive use of fossil fuel. A Conference of Parties to this Treaty (COP) is held every year and this was the catalyst for the renewed interest in nuclear power.

Providing the energy to sustain our way of life, in a planet with 8 billion people, was the problem; to move away from fossil fuel required alternatives, not only to replace our current use but to provide the additional energy that the developing countries needed to catch up. Global Energy requirements were forecast to increase by 30% between 2020 and 2050. In the world of non-fossil fuel, the Three Gorges Dam in China has a power capacity of 22,400 MW, the equivalent of 22 large nuclear power stations, and the wind turbines that are being built offshore, where size is not a problem, now have blades that exceed 100 metres, longer than a football pitch. Nevertheless, we are going to need much more non-fossil fuel than we can produce from these alternatives and a large growth in nuclear energy is a must if we are going to meet our commitment to carbon neutral by 2050.

By the time the twenty-sixth meeting of the UN Convention on Climate Change, COP26, came to Glasgow in November 2021, the birth of the second nuclear age had started to emerge, and it was COP26 that prompted me to write this book. I could see a rapid global growth in the nuclear industry with engineers, scientists, administrators, and accountants, as well as all the staff supporting the industry, wanting to understand the world of nuclear reactors. I also knew that explaining nuclear technology to the man in the street would not be easy, but I believed it could be done in a non-mathematical way. They would also be interested to know about all the peripheral aspects of nuclear power. Safety was top of the list, but many other new topics were starting to appear in the national press. Could hydrogen really replace the natural gas that heated our homes, and was it possible to actually reduce the atmospheric CO_2, the root of the problem, by capturing the carbon dioxide and storing it safely, and permanently, underground. My subtitle came along as an afterthought, when I realized that I was in fact telling the full story of the crisis we are facing with global warming and all the options for non-fossil fuel, not just nuclear. The hydrogen strategy, in particular, was something that needed to be explained.

This book gives an essentially non-mathematical account of all the background physics needed to understand what is going on inside a reactor. There had to be compromises, so I do use the equal sign, =, from time to time. I have tried to explain the physics with the help of diagrams and by writing in an informal manner, as though I am speaking face-to-face with the reader. I dilute the physics by telling quite a few anecdotal stories of happenings that took place during the development and understanding of the physics. The lexicon of physics needed to be explained, so the basic physics is covered in Chapter 2, and I put all the difficult physics in a separate chapter on Basic Quantum Theory, which you can ignore if you wish. I did include a separate chapter on relativity, since the man in the street sees it as way beyond him and I disagree; the origin of Einstein's famous equation $E = mc^2$ can be explained in plain language. Special attention is given to all the characteristics of the fission process and how nuclear reactor operators deal with their day-to-day problems in the control room. Safety is so important that a whole chapter is devoted to how the reactor designers and the regulators make sure that all possible risks have been taken into account, with solutions to deal with every eventuality. Even those scenarios that are difficult to accept, such as a terrorist attack, can still be managed safely and reactors with secondary containment can prevent radioactivity from escaping into the environment no matter what happens. The public have an understandable fear of nuclear radiation so I have used a set of stories to illustrate just how much that fear can be rationalized into an acceptance that nuclear radiation is everywhere and not of such great concern as they thought.

After dealing with the understanding of nuclear reactors I look to the future, but I don't have a crystal ball. I have attempted to forecast how nuclear reactors will develop as Generation IV plants start to appear. There are so many exciting developments of new technologies, but many of the proposals are taking a big leap forward, with ambitious goals, and some of them may prove to be a bridge too far.

Energy in the twenty-first century has two topics of peripheral interest. The first is nuclear fusion, which has been under development for over 70 years, but we still seem to be making slow progress, so I devote a chapter to explaining the problems that confront fusion and review the present status of fusion technology. The second topic is the hydrogen strategy. When the term 'hydrogen strategy' started to appear in the press I had difficulty understanding what it meant, and how it would fit in to our economic life. I think I can now see it as a possible revolution, waiting to take off, but commercial incentives will be needed to help it on its way. My chapter on hydrogen attempts to explain how this subtle concept will help us in our quest to achieve a net zero CO_2 lifestyle. However, we have to face the fact that the domination of fossil fuels as a source of energy is going to be difficult to replace, and I think it will still be the major global energy source for many decades.

When it comes to acknowledgements I must go back to my days at Stand Grammar School in Whitefield, a suburb of Manchester. The physics master, Les Lumley, was so motivating, that he unconsciously moved my choice of career away from chemical engineering into basic physics. I was also intrigued by nuclear physics, with radioactivity in particular, since my mother was having radiation treatment for cancer at the Holt Radium Institute. It was the mid-1950s when nuclear fission and Atoms for Peace were predicted to be the salvation of the world. I was enticed into nuclear physics and never looked back.

Writing this book has been a joy; most of the information was in my head, accumulated over many years in the UK Atomic Energy Authority. There are so many to thank for helping me to understand all the physics behind reactors, not only in the UK but in Canada, at Chalk River, and the US at Los Alamos. In recent years, since retirement, I have stayed in touch with quite a few colleagues through twice-a-year reunions, keeping up to date with all the developments in science and technology. I would like to thank all of my colleagues, but give a very special mention to John Leake, who provided many useful comments on the early drafts of every chapter. Richard Charles, Neil Jarvis, Eoin Lees, and Alwyn Langsford have all made a contribution to the book without realizing it.

My journey into the world of publishing has been a bit like climbing Everest, not quite as strenuous, but full of surprises. I have always had a bit of a problem with hyphenated words, and I didn't realize that hyphens came in three sizes: the standard, the longer, and the very long. Fortunately, I had the benefit of two very experienced Senior Editors, Sonke Adlung and David Lipp. They gave very polite answers to my stupid questions, and I thank them for guiding me to the summit. I am very grateful to Alex Greetham, a professional graphics artist and family friend. He stepped in at the eleventh hour and produced his own superb drawings for several internet images that I had intended to use but had difficulty in obtaining copyright consent for.

The final acknowledgement is traditionally family. An acknowledgement and thanks that can never match what they deserve. My wife, Val, has supported and encouraged me throughout, as have my children, Andrew, Suzanne, and Glenn. Giving details would be embarrassing, so I will leave them to simply read the fact, in this Preface, that the book is dedicated to, My Family.

Brian Hooton
July 2023

Contents

List of figures   xiii
Acronyms   xv

1. Introduction and Prelude   1
    1.1 Global Warming   1
    1.2 Capacity Factors   3
    1.3 Welcome to the Nuclear Age   5
    1.4 The First Electricity Producing Reactors   6
    1.5 The Prelude   8

2. Fundamental Nuclear Physics   11
    2.1 The Pauli Exclusion Principle   11
    2.2 Nuclear Forces   11
    2.3 Nuclear Reactions   12
    2.4 Energy and Mass Units   14
    2.5 Photons   14
    2.6 Antimatter, Pair Production, and Annihilation   16
    2.7 Mass Defects, Q Values, and Cross-sections   17
    2.8 Cross-sections   18
    2.9 The Discovery of Radioactivity   20
    2.10 The General Characteristics of Radioactivity   22
    2.11 Gamma Decay   23
    2.12 Spontaneous Fission   23

3. Basic Quantum Theory   24
    3.1 Skip This Chapter If You Wish   24
    3.2 The Uncertainty Principle   24
    3.3 The Theoretical Treatment of Nuclear Physics   25
    3.4 Atomic Spectra and Quantum Numbers   27
    3.5 Sommerfeld's Contribution   28
    3.6 Pauli's Contribution   28
    3.7 Spin and Parity   29
    3.8 Alpha Decay   30
    3.9 Beta Decay and the Story of the Neutrino   31
    3.10 The Discovery of the Neutron   32
    3.11 Quantum Theory and Beyond   33

4. The Story of $E = mc^2$ and Relativity   36
    4.1 The Unification of Electricity and Magnetism   36
    4.2 Relative Motion   37
    4.3 Einstein's Theory   38
    4.4 Standards of Mass, Length, and Time   40

5. The Fission Process and the Characteristics of Fission   41
    5.1 The Discovery of Fission   41
    5.2 Niels Bohr and Copenhagen   42
    5.3 The Fission Process   43
    5.4 Neutron Interactions   46
    5.5 The Fate of Gamma Rays   47
    5.6 Fission Fragments   48
    5.7 Delayed Neutrons   48
    5.8 The Energy of Fission   49
    5.9 Decay Heat   50
    5.10 The Chain Reaction   50

6. Nuclear Reactors in General   52
    6.1 Nuclear Reactor Calculations   52
    6.2 The Growth of the Neutron Population   52
    6.3 The Six Factor Formula   55
    6.4 The Effect of Delayed Neutrons on Reactor Control   56
    6.5 Reactivity   56
    6.6 Monte Carlo Models   57
    6.7 Nuclear Reactor Operations   58
    6.8 Fuel   59
    6.9 Moderators   60
    6.10 Coolants   61
    6.11 Poisons   62
    6.12 Control Poisons   62
    6.13 Unavoidable Poisons   63
    6.14 Burnable Poisons   65
    6.15 Engineering Materials   65
    6.16 The Fast Reactor   66
    6.17 Hybrid Reactors   68

7. Reactor Operations and Control   69
    7.1 Controlling Reactors to Keep Them Safe   69
    7.2 The First Reactors   69
    7.3 Reactor CP1   70
    7.4 Controlling Commercial Reactors   71
    7.5 The Reactor Pressure Vessel   72
    7.6 The Reactor Coolant Pump   73
    7.7 The Pressurizer   73
    7.8 The Steam Generator   73
    7.9 The Boron Loading Loop   74
    7.10 Power Measurement   75
    7.11 The Fuel Temperature Coefficient (FTC)   76
    7.12 The Moderator Temperature Coefficient (MTC)   77
    7.13 The Void Coefficient (VC)   78
    7.14 Changes in Steam Demand   78
    7.15 Control Room Operations   78

8. Safety   81
    8.1 Safety, Risk, and Consequences   81
    8.2 The Regulators   82

8.3 Decay Heat Removal 82
8.4 Loss of Coolant 83
8.5 Passive Safety Measures 84
8.6 The Windscale Fire 84
8.7 Brown's Ferry 85
8.8 Three Mile Island 87
8.9 Chernobyl 1986 88
8.10 Problems in the Fukushima Region of Japan 89
8.11 Safety Overview 90
8.12 Understanding the Health Hazard of Radiation 92

9. The Nuclear Fuel Cycle 97
9.1 The Nuclear Fuel Cycle Definition 97
9.2 Mining 98
9.3 Enrichment 98
9.4 Fuel Fabrication 99
9.5 Spent Fuel Management 100
9.6 Spent Fuel Ponds 102
9.7 Cherenkov Radiation 102
9.8 Reprocessing 102
9.9 Nuclear Waste 105

10. International Treaties and Obligations 107
10.1 Euratom 107
10.2 Treaty on the Non-Proliferation of Nuclear Weapons (NPT) 108
10.3 The International Atomic Energy Agency (IAEA) 108
10.4 Nuclear Safeguards 108
10.5 Obligations 110

11. The Future of Fission Reactors 111
11.1 The Alternatives to Fossil Fuel 111
11.2 Generation IV Technology 111
11.3 The Move to Higher Temperatures 112
11.4 The Move to Fast Reactors 112
11.5 The Move to Modular Reactors, SMRs, and AMRs 113
11.6 Plutonium Breeding 114
11.7 Thorium Breeding 115
11.8 New Coolants 116
11.9 Molten Salts 116
11.10 New Types of Fuel 118
11.11 Burning Waste and Using the Minor Actinides as Fuel 118
11.12 New Reprocessing Technology 118
11.13 The Economics and Politics of Electricity Generation 119
11.14 The Utilization of $E = mc^2$ 121

12. Nuclear Fusion 122
12.1 The Fusion Process 122
12.2 Producing Fusion in the Laboratory 123
12.3 ITER 125
12.4 MAST and STEP 125
12.5 The Fuel for Fusion 126

12.6 The Tritium Breeding Ratio (TBR)   127
12.7 Venture Capital   129
12.8 The Conclusion on Fusion   129

13. The Hydrogen Strategy   131
13.1 The Basic Properties of Hydrogen   131
13.2 The Production of Hydrogen   132
13.3 Carbon Capture   135
13.4 Energy Storage   136
13.5 New Markets for Hydrogen   136
13.6 Hydrogen in the Colours of the Rainbow   138
13.7 The Race to Deliver Net Zero   138

Further Reading   140
Index   146

List of figures

Figure 1.1. CO_2 levels in the atmosphere   1

Figure 1.2. The first nuclear age. The twentieth-century growth and decline of nuclear reactors   4

Figure 2.1. The hook and eye interpretation of the short-range nuclear force   12

Figure 2.2a. Proton repulsion and scattering due to positive electric charge   13

Figure 2.2b. A proton unable to penetrate the nucleus   13

Figure 2.2c. The path of a proton with sufficient energy to get over the Coulomb barrier. It becomes captured in the nuclear potential well by the strong nuclear force   13

Figure 2.3. The definition of the electron volt as a unit of energy. The electron has V electron volts of energy after being accelerated across an electrical potential of V volts   14

Figure 2.4. Newtons rings. An interference pattern, illustrating the wave-like nature of light   15

Figure 2.5a. Pair production   16

Figure 2.5b. Electron-positron annihilation   17

Figure 2.6. An energy level diagram depicting the alpha decay of U^{238} 18

Figure 2.7. Neutrons incident on a slab geometry   19

Figure 2.8. The U^{238} decay chain   21

Figure 3.1. Showing discrete line spectra when electrons fall into lower energy levels   27

Figure 3.2. The path of an alpha particle escaping from the strong nuclear forces in the potential well by tunnelling through the Coulomb barrier   30

Figure 3.3. A particle depicted as a travelling wave packet   34

Figure 4.1. What is the meaning of simultaneous?   38

Figure 4.2. Two coordinate systems, B moving at a constant velocity relative to A   39

Figure 5.1. The liquid drop model of fission. After the nucleus absorbs a neutron, it distorts into the shape of a liquid drop. Fission takes place with a release of energy and neutrons   41

Figure 5.2. The energy spectrum of neutrons from fission. The shape is given by an empirical fit of the data to a mathematical expression known as the Watt spectrum   44

Figure 5.3. Typical Maxwell distribution of velocities   44

Figure 5.4. The U^{235} fission cross-section. The resonance region is clearly shown   45

Figure 5.5. Percentage energy loss by a neutron in a head on collision with various nuclei   46

Figure 5.6. Approximate thermal neutron absorption cross-sections for various nuclei. Cross-section in barns   46

Figure 5.7. Compton scattering, a billiard-ball type of collision between a photon and an electron. The scattered photon has a longer wavelength and smaller frequency. It has less energy, lost to the electron recoil 47

Figure 5.8. The fission product mass distribution from thermal neutron fission in U^{235} 48

Figure 5.9. Details of the six delayed neutron groups. More than one radioactive precursor may contribute to each group 49

Figure 5.10. Nuclear decay heat reduction as a function of time after shut-down 50

Figure 5.11. Fission cross-sections for thermal and 2.0 MeV neutrons. Numbers of prompt and delayed neutrons from fission by thermal and 2.0 MeV neutrons 51

Figure 6.1. A selection of poisons. Thermal (Maxwellian) capture cross-sections; barns 62

Figure 6.2. The Xe^{135} decay chain 63

Figure 6.3. Time dependence for residual Xe^{135}. The initial rise is due to the formation of Xe^{135} from the decay of I^{135}. The right-hand axis shows the effect of Xe^{135} on reactivity 64

Figure 6.4. The Sm^{149} decay chain 65

Figure 7.1. The main components of a PWR 72

Figure 7.2. A PWR pressurizer. The bottom heaters and the top spray nozzle are used to control the pressure 74

Figure 7.3. A steam generator. Chevron and swirl-vane separators are used for steam drying 74

Figure 7.4. U^{238} capture cross-section 77

Figure 9.1. The nuclear fuel cycle 97

Figure 9.2. Uranium mining to produce yellow cake 99

Figure 9.3. Uranium hexafluoride phase diagram 100

Figure 9.4. A uranium enrichment centrifuge 101

Figure 9.5. The PUREX process 104

Figure 9.6. A mixer-settler 104

Figure 12.1. Fusion cross-sections. Since a temperature covers the Maxwellian spread of velocities, the cross section is given as an average over the velocity spread 123

Figure 12.2. Venture capital alternatives to conventional Tokamak fusion 130

Figure 13.1. The calorific value of various fuels 132

Figure 13.2. An illustration of electrolysis. It is used to split water into hydrogen and oxygen 133

Figure 13.3. Hydrogen production costs. From the International Energy Agency, 2019 134

Acronyms

AGR	Advanced gas-cooled reactor
amu	atomic mass unit
BWR	Boiling water reactor
CANDU	Canada deuterium uranium
CCFE	Culham Centre for Fusion Energy
CCS	Carbon capture and storage
CCU	Carbon capture and utilization
CND	Campaign for Nuclear Disarmament
COLEX	Column exchange, lithium enrichment
COP	Conference of Parties to the UNFCCC
DAC	Design acceptance confirmation
DEMO	Demonstration of electricity from fusion
EAST	Chinese Experimental Advanced Superconducting Tokamak
FTC	Fuel temperature coefficient
GLEEP	Graphite low energy experimental pile
GNEP	Global Nuclear Energy Partnership
HEU	High enriched uranium
HLW	High-level waste
HTGR	High-temperature gas-cooled reactor
IAEA	The International Atomic Energy Agency
IDDP	Iceland Deep Drilling Project
IEA	International Energy Agency
INES	International Nuclear and Radiological Event Scale
ITER	International thermonuclear experimental reactor
JET	Joint European Torus
KSTAR	Korea Superconducting Tokamak Advanced Research
LCOE	Levelized cost of electricity
LEU	Low enriched uranium
LLW	Low-level waste
LOCA	Loss of coolant accident
MAST	Mega ampere spherical tokamak
MASTU	MAST upgrade
MDEP	Multinational design evaluation programme
MOX	Mixed oxide
MSR	Molten salt reactor
MTC	Moderator temperature coefficient
NEA	Nuclear Energy Agency
NPT	Treaty on the Non-Proliferation of Nuclear Weapons
NRC	Nuclear Regulatory Commission
NRPB	National Radiological Protection Board
PEM	Polymer electrolyte membrane
PFR	Prototype fast reactor
PORV	Pilot-operated relief valve
PUREX	Plutonium uranium extraction
PV	Photo voltaic
PWR	Pressurized water reactor
RCS	Reactor control system
RCS	Reactor coolant system
RHS	Residual heat removal system

RPS	Reactor protection system
SC	Secondary containment
SFM	Special fissile materials
SI	Safety injection system
SMR	Small modular reactor
SMR	Steam methane reforming
SoDA	Statement of design acceptability
SOEC	Solid oxide electrolyte cell
SUR	Start-up rate
TBR	Tritium breeding ratio
THORP	The UK Thermal Oxide Reprocessing Plant
TMI	Three Mile Island
TRISO	TRistructural-ISOtropic Fuel
UNDO	Un-do CO_2
UNFCCC	United Nations Framework Convention on Climate Change
VC	Void coefficient

1

Introduction and Prelude

1.1 Global Warming

Global warming is starting to bite, with record temperatures, year after year, and more intense hurricanes, forest fires, floods, and droughts. It was a while before carbon dioxide was universally accepted as the villain, but the current levels of CO_2 are much higher than they have ever been. Figure 1.1 shows the atmospheric CO_2 levels as we pass through the ice ages fluctuating up and down like a yo-yo, but never higher than 300 parts per million (ppm). The Industrial Revolution was the beginning of a slow growth in CO_2 levels; it kept on accelerating and has achieved a height of 420 ppm in 2023. Each year nations gather at the COP conferences to thrash out a solution. COP is the Conference of Parties to the UN Framework Convention on Climate Change (UNFCCC), a treaty established in 1994 with 165 signatories. It is of course a global problem that no nation can solve in isolation, and with 8 billion inhabitants, the planet has a massive problem to deal with.

The problem of global warming is not limited to extreme weather conditions since the melting at the poles will result in a rise in the sea level, but we are not sure how much it will rise, since we don't yet have the scientific data to give reliable estimates. We should have better sea-level-rise data during the 2030s, but it could turn out to be even more alarming than the change in our weather. If all the ice in the polar regions melts, we know for sure that the sea level would rise by about 100 m. Unreliable estimates abound and some suggest that a 2 m rise by the end of the century may happen, which would be a real disaster for many islands and coastal cities. If we are

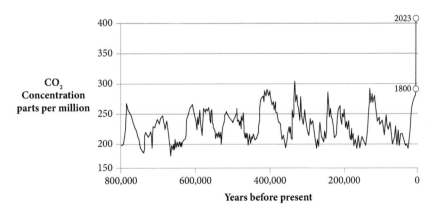

Figure 1.1 CO_2 levels in the atmosphere

Source: Darrin Qualman

Understanding Nuclear Reactors. Brian Hooton, Oxford University Press. © Brian Hooton (2024).
DOI: 10.1093/oso/9780198902652.003.0001

able to reduce the CO_2 in the future then our weather patterns would quickly return to normal, but not the sea level. Getting the level of the oceans down requires waiting for snow to fall and remain in the polar regions, and this will take thousands of years.

The root cause of the problem is energy, since it dominates our lives. We need it for all our modes of transport, trains, jet planes, boats, and automobiles, electric or otherwise. We need it to live in warm/cool comfort with air-conditioning, and to provide the energy for cooking, lighting, and our entertainment, not to mention the ice cubes in our cocktails! An enormous amount of energy is needed to manufacture the essentials of our civilization, all the goods we are not prepared to do without. Everything we can see, and touch, has energy associated with it, even the paint on the walls.

It would be great if we took the problem of global warming more seriously. If we treated it as a threat similar to a war, and moved onto a war footing as we did in 1939 by abandoning the normal constraints of peacetime, we could give the problem the immediate attention it deserves. In 1941, the USA fleet in the Pacific, at the time of Pearl Harbour, had just four aircraft carriers but, by 1945, with the benefit of emergency powers and the cash from War Bonds, they had 40 aircraft carriers—they were building one a month. Similar remarkable efforts were made by Britain, Russia, and even Hitler's Germany. Unfortunately, our democracies are unlikely to be prepared to move onto a war footing to deal with global warming.

Having painted such a bleak picture of the future as a result of global warming, I need to cheer myself up with a few optimistic scenarios. Since the beginning of the Industrial Revolution, most of our energy has come from fossil fuel, coal, oil, and natural gas, but to avoid releasing CO_2 into the atmosphere we must find alternatives, and historically we have several very useful starting points. Centuries ago, windmills were used to grind corn and they have now developed into wind turbines that generate electricity. The gigantic windmills that have three blades and are being built in the shallow waters of the North Sea, just off the east coast of England, are amazing. Each blade is over 100 metres long, more than the length of a football pitch, and each turbine is now capable of delivering more than 14 megawatts (MW) of power. The Aswan Dam in Egypt (2,100 MW), built in the 1960s, and the Kariba Dam (1,626 MW) in central Africa, completed in 1959, have been producing large quantities of hydroelectricity for many years. The more recent Three Gorges Dam in China, the largest dam in the world, is capable of delivering an enormous 22,500 MW of hydroelectric power, equivalent to 22 large nuclear power stations.

Solar panels convert sunlight into electricity by the photovoltaic process, PV. It converts light photons into volts. The process was originally discovered by Edmond Becquerel as far back as 1893 and gradually became commercially viable with the modern design of solar cells emerging in 1939. They continue to improve in efficiency and the installation cost has dropped by 90% in the last 10 years, making them now predicted to become the cheapest form of electricity in history, according to the

International Energy Agency (IEA). In the long term, we hope to have completed new technology to mount lightweight solar panels on satellites, and beam the electricity back to Earth—blue sky research for real.

There are other renewable technologies, with several methods for harvesting the wave and tidal energy already helping the quest for net zero. Geothermal energy has been an optimistic hope for decades, but it does have a major problem. It is possible to get high-temperature gas or steam out of the hot zone at the bottom of deep bore holes but the very act of removing it cools the hot zone and it can take much too long for the hot zone to recover. Iceland is a well-known example of geothermal energy but although the current boreholes, at a depth of 2.5 km, can provide low-grade heat for heating domestic dwellings and some electricity, the Iceland Deep Drilling Project (IDDP) is exploring down to a depth of 5 km in the hope of extracting water at 500°C.

The words 'energy' and 'power' are used so frequently in general conversation that it is important to understand the distinction. Power is the ability to produce energy. We may well have a 3 kW electric kettle, but it doesn't create energy until it is switched on, and it stops creating energy when it is switched off. If a 3 kW kettle is switched on for 20 mins, the energy produced is:

3.0 × (a third of an hour) = 1.0 kilowatt hour, kwh; these are the energy units that we are familiar with, and appear on our electric bills.

The specification for the energy of a nuclear reactor needs to be clear about the distinction between thermal power and electricity generated. A 500 MW (e) electrical reactor would have a thermal rating in the region of 1,500 MW (th) if the efficiency for the conversion of heat into electricity is 33%.

1.2 Capacity Factors

When a discussion on wind turbines takes place, someone always says—but the wind isn't blowing all the time! and in the case of solar panels we know the sun doesn't shine at night. The solution to this problem is to state exactly how much energy was produced in a particular year. This is known as the **capacity factor**. It is a historic, actual figure, for a particular power plant in a particular year, and an example is the capacity factor for a wind turbine in the North Sea of 42% in 2017, it was a windy year. If a wind turbine has a capability of producing 1 MW (The Plate Label), then during the whole year of 8,760 hours it could produce 8.76×10^6 kWh of electricity. If, in a particular year it only produced 3.67×10^6 kWh then the capacity factor for that year is given by:

$$(3.67 \times 10^6)/(8.76 \times 10^6) = 0.42$$

Capacity factors can be applied to all forms of energy production. The Three Gorges Dam in China had a year when the capacity factor was only 34%, maybe due to drought or turbines being out of action, or simply by closing the valves at night time when the demand for electricity was low. Some interesting examples of capacity factors are given as follows:

Wind turbines—43%
Solar panels—UK and most of Europe—10%
Solar panels—Arizona—20%
Solar panels—Australia—29%
Three Gorges Dam—34%
Nuclear power station—91%

All of these alternatives have a big role to play in the battle against global warming, but they are still well short of meeting global demand. The nuclear option is well capable of generating very large amounts of energy. The first nuclear age came to an end following the disaster at Chernobyl in 1986. No politician would dare to put the word 'nuclear' on any agenda; it was a vote loser. Instead, the nuclear capacity was allowed to drop and, as we see in Figure 1.2, the first nuclear age came to an end in 1994 when there were zero new constructions. Global warming has put nuclear power back on the agenda and given birth to the second nuclear age.

This book anticipates a large growth in the nuclear industry and is intended, as the title suggests, to give a good understanding of nuclear fission reactors, their safety, and their ability to help the battle against global warming. I begin with a recap of the technology that emerged through the first three generations of nuclear power and a brief historical account of the Manhattan Project.

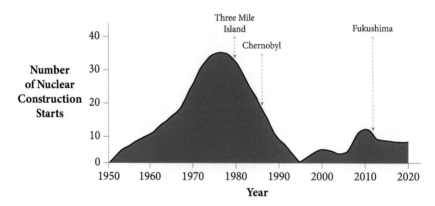

Figure 1.2 The first nuclear age. The twentieth-century growth and decline of nuclear reactors

1.3 Welcome to the Nuclear Age

The Four Generations of Nuclear Reactors

As we moved into the twenty-first century, we began to develop what are referred to as Generation IV reactors. Previous generations did not have precise definitions, but the retrospective view is along the lines of the criteria given as follows.

Generation I

The theme of the first generation was proof of concept (i.e. we have done the calculations, so let's build it and see if it works!) Most of the reactors used natural uranium since enrichment technology was in its infancy, but one of the first power reactors, Shippingport, USA, was very much a test bed for learning. It used various fuels in its operations, starting with a mixture of U^{235} enriched to 93% and natural uranium. Calder Hall in the UK was a prototype in this category using natural uranium.

Generation II

After the proof of the scientific and engineering principles had shown nuclear power would be technically viable, a host of designs were put forward using variations on the following themes,

Natural or enriched uranium in the form of metal or oxide.
Graphite, normal water, or heavy water as a moderator.
Water or gas as a coolant.

The emphasis was now on cost, safety, and commercial viability.

The reactors that successfully emerged from this period were the boiling water reactor (BWR) and the pressurized water reactor (PWR). The Canadian reactor, using deuterium in the form of heavy water (CANDU), and the advanced gas-cooled reactors (AGR), also came into successful operation. Their operational lifetime was anticipated to be about 25 years, but the average has turned out to be about 40 years. They used traditional active safety features expecting powered emergency pumps to start automatically when needed. Equipment redundancy played a significant part in safety philosophy. In the USA, Generation II was more or less defined by the regulation NRC 10 CFR Part 50.

Generation III

Gen III reactors are similar to Gen II but with improved thermal efficiency, some modular construction, improved fuel design, standardized design, and the introduction of some passive safety systems. The only distinction that defines Gen III is that they became regulated in the USA by NRC 10 CFR Part 52.

Generation III+

This generation can hardly be described as distinctive, and I view it as the psychological answer to Chernobyl. Safety improvements in new designs were proposed, but still under USA 10 CFR Part 52.

Generation IV

At the turn of the century, nuclear reactors were still relying on Generation III plant designs, most of them PWRs, but new technology was about to emerge. Many of the new suggestions should be regarded as pioneering. They have distinct advantages over previous reactors but would require extensive R&D before becoming a commercial reality. The US started the ball rolling in 2000 when they formed the Generation IV International Forum (GIF). It was formally chartered in 2001 with the noble objectives of sharing technical knowledge, sharing ideas, and avoiding too much duplication. All the major players joined at some stage, including the USA, China, Russia, France, Japan, and the UK. The forum examined all the suggested technologies and arrived at a set of six, for shared development as Generation IV reactors. These will be covered later in the chapter on the Future of Fission (Chapter 11).

1.4 The First Electricity Producing Reactors

Before nuclear reactors started to deliver electricity the USA and the UK operated nuclear piles for the production of plutonium. They later used similar piles for the production of tritium, required for the development of the H bomb. In the UK the first reactor was the Graphite Low Energy Experimental Reactor, **GLEEP**, built at Harwell in 1947. The two plutonium piles were built at Windscale in 1950 and 1951. The USA's first pile, after the Fermi pile CP1, was the X 10 at Oak Ridge, Tennessee. It produced the first kilos of plutonium before the Handford piles became operational during the Manhattan Project. In October 1956, Queen Elizabeth II opened the world's first electricity-producing nuclear reactor, Calder Hall, in Cumbria, UK. It was the birth of Generation I. A second reactor of similar design, 60 (e), graphite-moderated, and carbon-dioxide cooled, was commissioned at Chapelcross in 1959.

In the USA, their first electricity-producing reactor was the SM-1 at Fort Belvoir, Virginia, in 1957. It was only 2 MW (e), and the electricity it produced was for the benefit of military facilities. The Shippingport reactor, 60 MW (e), started to produce electricity in 1957 and was the first reactor to use enriched uranium as a fuel. It used a mixture of natural uranium and uranium enriched to 93%. In the 1950s, natural uranium was readily available as a fuel, with graphite the obvious moderator since Fermi had shown that the combination worked with his very first critical pile in Chicago. Eventually, enriched uranium became available, which opened the door to a multitude of options for fuel, moderator, and coolant.

Heavy water-moderated reactors were possible with natural uranium but supplies of heavy water were limited and it was realized that ordinary water could be used both as a moderator and a coolant if the fuel was slightly enriched in U^{235}. Alternatives to water were suggested, with molten salts and gases, such as CO_2 and helium, being tested in experimental reactors. In 1965 at Oak Ridge, USA, the Molten Salt Reactor Experiment used a mixture of fluoride salts as a coolant and tested many variations, including the use of U^{233} as a fuel dissolved in the coolant itself. These were the days of innovation and, since there was some fear, at this time, that oil wells would run dry, the decision to build experimental breeder reactors soon became a common objective in the USA, Europe, and Russia.

High-temperature gas-cooled reactors, **HTGRs** were operated as demonstration plants in Germany, the UK, and the USA, each for about 10 years. The attraction of high temperature was the improvement in thermodynamic efficiency, which was limited to about 35% using steam at 315°C. The notion that temperatures above 800°C would be useful for processing heat and hydrogen production was not yet apparent. The demonstration plants also provided the framework to test various fuel types, such as ceramic pebbles clustered together in a cylinder—the pebble bed fuel concept. Although many of these demonstration plants did not lead to any significant commercial exploitation they provided valuable experience for the twenty-first-century developments known as Generation IV. Natural water became the favourite medium for commercial reactors, with the PWR and the BWR becoming the most popular choice. The Heavy Water Reactor, developed in Canada, CANDU, had notable success and is still a favourite choice for many.

Calder Hall was designed to produce plutonium for the UK nuclear weapons programme, but it was also the beginning of a surge in the building of nuclear reactors for peaceful purposes. President Eisenhower proclaimed a doctrine of Atoms for Peace, in an address to the United Nations in 1945. It supplied equipment and technology to emphasize the peaceful opportunities of nuclear energy. Two of the beneficiaries were Israel and Pakistan. Ironically, both now recognized nuclear weapon states.

Oil supplies were forecast to diminish, and maybe even disappear, by the end of the century, and this generated a fear that there might not be enough uranium to go around. With this possibility at the forefront of political thinking Euratom had been set up in 1957 to ensure that uranium would be shared equally amongst its members. The climate changed with the discovery of North Sea oil in the early 1960s and with the Campaign for Nuclear Disarmament (CND) membership growing, all it would need for the civil nuclear programme to collapse would be another nuclear disaster. The Windscale fire occurred in 1957 at one of the military nuclear piles on the same site as Calder Hall. It had the population reeling with media hype about radioactive iodine in milk. Sure enough, in 1979 Three Mile Island (TMI) became the public relations champion of the CND and, to cap it all, Chernobyl in 1986 brought the first nuclear age to an end.

As the twentieth century came to an end it seemed that oil supplies would never run out, with ever more oil and gas fields being discovered as well as new technologies to extract oil from oil shale. Fracking came a little later but there was no incentive for any politician to put anything with the word 'nuclear' on the agenda. Elections were every four years and mentioning nuclear would definitely lose votes. There was no need for nuclear, so what could bring about the beginning of the second nuclear age? The catalyst for the second nuclear age was the Intergovernmental Panel on Climate Change, established in 1988. Scientists were divided on the cause of global warming with many well-respected professors claiming it was just the aftermath of an ice age, modified by statistical variations. However, slowly but surely, opinions merged into the now well accepted conclusion that climate change is due to the build-up of CO_2 in the atmosphere, and that this is caused by burning fossil fuels. End of the story, the second nuclear age was in the incubator. The growth and decline of nuclear power in the twentieth century are illustrated in Figure 1.2.

The obvious and main alternatives to fossil fuels are solar panels, wind, and nuclear. They will all be needed, and since the planet has vast open spaces where the sun shines, like the Sahara and the vast interior of Australia, we have ample opportunities, and space, to harvest solar energy and turn it into hydrogen. Nuclear is going to play a big role into moving away from fossil fuels, even the politicians now openly support it. It can provide large quantities, gigawatts 24/7, and only use a fraction of the space that the other options require. The nuclear option will be an essential part of our fight to deal with global warming. We have travelled through exciting times in the first nuclear age, a journey that began during the Second World War with the establishment of the first chain reaction, and the amazing achievements of the Manhattan Project. The economic and political aspects of all these competing alternatives to fossil fuel is considered at length in Chapter 11, the Future of Fission.

1.5 The Prelude

The story of nuclear reactors has a prelude, and you will recognize it as the well-known Manhattan Project, the race to build an atomic bomb in the USA before Hitler could do the same. Although the fission process had been discovered in December 1938, there were serious gaps in our knowledge and they would not be filled overnight. It was not until 1942, when the Manhattan Project got off the ground, and Enrico Fermi's CP 1 nuclear pile showed that a sustained chain reaction was possible, that we realized that we could entertain an optimistic view on the future of nuclear energy. The events of the next three years would fill all the gaps in our knowledge and set the scene for the eventual use of nuclear energy for the benefit of all mankind; Atoms for Peace.

Brigadier General Leslie Groves was appointed as director of the Manhattan Project in September 1942. Who was he? He was an officer in the US Army Corp of Engineers who had just finished building the Pentagon and was looking forward to a well-earned rest. His training in the Corp of Engineers was relevant since the Corp had built the Panama Canal, way back in 1907, and had learned the valuable lesson that the first action on any major project is to take good care of the work force. The French attempt to build the canal had failed because yellow fever and malaria killed many of the work force. When the first USA reactors were built at Hanford in a remote part of Washington state, to produce plutonium for the bomb, the first priority was to build the roads, railways, power plants, accommodation, and a baseball diamond. Groves appointed Robert Oppenheimer as the scientific director, but Leslie Groves was the man with the money, and although Pearl Harbour was a blank cheque for funding, the secret project still required approval by Congress, quite a problem to ask Congress for a small fortune without saying what the money was for! President Roosevelt used his political tact to overcome this problem by wearing his hat as Chief of the Armed Forces, which meant he could spend what he wanted, without Congressional approval.

There were two major projects to provide fissile material in a quality and quantity fit to make a bomb. The first was to establish new technology for the enrichment of uranium. It had turned out that the uranium isotope required for a bomb was U^{235}. Quite a disappointment since it was only 0.7% of natural uranium. Even so, it turned out to be enough to create a chain reaction with natural uranium, but for a bomb, a high enrichment, greater than 90%, was the challenge. One of several methods to produce high enriched uranium was to use magnetic separation and a massive plant was built at Oak Ridge. The magnets required some 5,000 tonnes of copper for the windings, and it was in short supply, but Groves was not prepared to wait, and demanded 6,000 tonnes of silver from the Treasury Bullion Depositary. His request solicited the response: 'You may think of silver in tonnes, but the Treasury will always think of silver in troy ounces'. In the end the project used over 13,000 tonnes of silver and more than 96% was recovered and returned to the Treasury at the end of the war. The other fissile material that could be used for a bomb was plutonium. It had only been discovered in 1940 but it was soon recognized as a suitable fissile material and had the benefit that no isotopic enrichment would be needed. Plutonium was a natural by-product of a uranium-fuelled reactor through neutron capture in U^{238}. This led to Np^{239} and then, after radioactive decay, to Pu^{239}. The spent fuel would require reprocessing to separate plutonium from all the other elements, the approval to build a reactor and reprocessing plant at Hanford for the production of plutonium was given in 1943. It was considered to be a hazardous activity, so the employee village had to be not less than 10 miles upwind of the plant! The main reactor, reactor B, was constructed on a green field site in the middle of nowhere, and with the benefit of the emergency conditions of the war the construction was completed in just 13 months!

The main scientific centre for the Manhattan Project was Los Alamos. This was where theoretical work, and experiments to determine all the nuclear parameters needed to design the bombs, would be carried out. The uranium bomb was not a problem. The physics showed that a bomb would work by firing two sub-critical masses together and, although it required a very high speed to get an efficient release of energy, modern artillery guns could do the job. The plutonium bomb was a different kettle of fish. It was decided that, to prevent a pre-detonation, a new technology, an implosion device, had to be developed. This consists of segments of plutonium surrounded by a complete sphere of conventional explosives. The conventional explosives had to be detonated simultaneously and would implode, compressing the plutonium segments into a super-critical mass, a very effective bomb.

The uranium bomb was code named 'Little Boy' and the plutonium bomb, which was much larger, was called 'Fat Man'. Both were deliverable by a B57 bomber, but an engineering modification was required to accommodate the much larger Fat Man. The engineers who were working on this very highly classified 'Fat Man Modification' thought it was for Winston Churchill!

The Manhattan Project had called upon many of the UK's scientists to participate, at the expense of several other areas of war technology crying out for expertise. The code breakers at Bletchley Park had high priority as did the development of radar, but many of the very best brains sheltered behind their ivory towers at prominent universities. The professor of physics at Cambridge fought a vigorous battle to persuade one of his staff to join the radar team. He was faced with the objection that—'to work on radar all you need to know is Ohm's Law'. The professor responded with—'yes, I agree, but you have to know it really well!'

The two bombs were a successful, but horrific, demonstration of the power of nuclear energy. The images of Hiroshima and Nagasaki became the publicity leaflet for the 'Ban the Bomb' movement that still flourishes today. Philosophically, If H and N had not happened, we would not be quite so aware of how awful the prospect of a nuclear war would be and perhaps, because we did see the reality of atomic bombs, in a vision permanently etched in our minds, we have managed to avoid nuclear conflict, so far!

The stage was now set for the development of nuclear energy for peaceful purposes, the prelude was over, and nuclear data were now available to enable the building of commercial, electricity-producing, reactors.

2

Fundamental Nuclear Physics

2.1 The Pauli Exclusion Principle

Most of us have a warm understanding about the structure of the atom. We see it as a positively charged nucleus with negatively charged electrons orbiting around it, just like the planets orbit the sun. The planets are kept in position by gravitational forces and the electrons orbit the nucleus using the electric attraction of the negatively charged electrons to the positively charged nucleus. The orbits in the atom, often called shells, are restricted by the rules of modern physics, the acceptable orbits for electrons have identifying characteristics called quantum numbers. You can think of them as a unique address, where the electrons live. There is a restriction, that only one electron can live at any address. This is referred to as the **Pauli Exclusion Principle,** which he proposed in 1925. The Principle has a wider application than electrons, and applies to all fermions, a class of elementary particle with odd half-integer spin, such as ½, or 3/2, for example.

Electrons can be kicked out of their orbit by energy supplied from various sources, electric discharges for example, and we know that electrons jumping between the outer orbits of an atom create visible light. The inner orbits, on the other hand, are much more energetic, they have identifying letters, the K shell being the innermost shell, followed by the L shell, etc. Electrons ejected from these inner shells are rapidly filled by electrons dropping down from higher shells and releasing energy in the form of photons, with much higher energy than visible light, the inner shell vacancies create X-rays.

If we delve into the core at the centre of the atom, we find the nucleus contains two types of particle, the positively charged proton and the neutron, with no charge. Both of these are referred to as **nucleons**, particles that exist in the nucleus; they are about 1,800 times heavier than an electron. If we were to delve even deeper into the interior of these nucleons, we would enter the world of quarks and other fundamental particles in our universe. However, we don't need to go any deeper than the two nucleons, the proton, and the neutron, to understand the processes of nuclear physics and what makes nuclear reactors tick.

2.2 Nuclear Forces

Protons are positively charged and should repel each other, so what holds the nucleus together? The nucleus is stable because a different force in nature, the **strong nuclear force**, overcomes the repulsive electrical force. The nuclear force is quite different from

Understanding Nuclear Reactors. Brian Hooton, Oxford University Press. © Brian Hooton (2024).
DOI: 10.1093/oso/9780198902652.003.0002

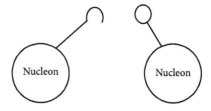

Figure 2.1 The hook and eye interpretation of the short-range nuclear force

electrical and gravitational forces that operate over any distance. The strong force operates over only a short range, in the region of 10^{-13} cm, and is quite complicated in its
detail. Gravitational and electrical forces extend forever, and it is difficult to visualize
a force that has a limited range. It is sometimes depicted as shown in Figure 2.1. Like
a hook and eye, that can link up at a certain distance but would have no effect beyond
the range of the hook. The strong nuclear force is referred to as strong because there
is yet another type of force, the weak nuclear force. We will encounter the weak force
later, when we explain beta decay in Section 3.9 on radioactivity. Let us now look at
what happens when a proton from a nuclear accelerator approaches a typical nucleus.

2.3 Nuclear Reactions

Figure 2.2a. shows a proton approaching the nucleus, and because of the electric force
of repulsion it gets deflected, scattered is the technical term, it goes off at an angle, with
the nucleus in recoil. In Figure 2.2b, which is a head-on collision, it slows down as it
approaches the nucleus and the force of repulsion will bring it to a stop. However, it
has potential energy, so it turns around and back-tracks along the way it came under
the repulsive force of the nucleus. It does not have sufficient energy to overcome the
electrical repulsion and get into the nucleus. If, however, the proton has enough energy
to get close to the surface of the nucleus, the nuclear force, which is a strong force
of attraction, can grab hold of it and pull it in to the centre of the nucleus, proton
capture. Figure 2.2c shows the path to capture a proton into the nuclear interior. As the
proton approaches from the right the electrical force of repulsion increases, but when
it reaches the surface, it encounters a very strong, attractive force and gets pulled down
into the centre of the nucleus. The region inside the nucleus is referred to as the **nuclear
potential well**, inside which strong nuclear forces hold the nucleons together. The high
point at CB is called the Coulomb barrier, the name Coulomb being the unit of charge
responsible for the electric force. In summary, the energy of the proton must be more
than the Coulomb barrier to get into the nucleus. This restriction does not apply to
neutrons, since they have no electrical charge.

   The internal structure of the nucleus is somewhat similar to the atom. The nucleons
are not in orbits around some central core, but they do occupy energy states, usually referred to as energy levels, with quantum mechanical rules, similar to those that
govern electrons in the atom. When the nucleus changes its energy level it may emit a

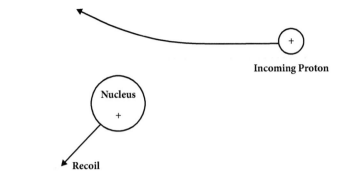

Figure 2.2a Proton repulsion and scattering due to positive electric charge

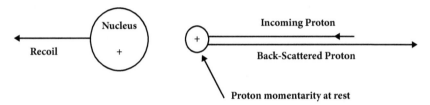

Figure 2.2b A proton unable to penetrate the nucleus

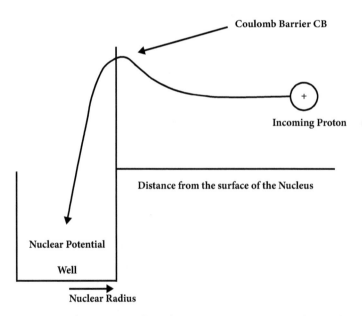

Figure 2.2c The path of a proton with sufficient energy to get over the Coulomb barrier. It becomes captured in the nuclear potential well by the strong nuclear force

photon, which is usually of higher energy than X-rays. These photons, originating from inside the nucleus, are called gamma rays. Gamma rays can have an energy greater than or less than an X-ray and both names, X-ray and gamma ray, should be viewed as referring to the place of origin, not the actual amount of energy.

2.4 Energy and Mass Units

Before we get into more detail about the subject of nuclear physics, we need to define the language we will use. The units of mass are not the kilogram, which is far too large for practical use, it is the **atomic mass unit, amu,** now given a name, the Dalton, Da. The Dalton is one-twelfth of the mass of the C^{12} atom, including its electrons. For some reason, the Dalton is not a popular name and the amu is still in common use.

The units of energy used in nuclear physics are also different from the kilowatt hour that we are used to in domestic living. The **electron volt, eV,** is a unit of energy equal to the energy acquired by an electron accelerated across a potential of 1 volt, see Figure 2.3. The energy of a proton accelerated across this voltage would be exactly the same because they have the same charge. However, since the proton is much heavier than the electron its velocity would be much smaller to give it the same energy. In practice, we will use **keV** for a thousand electron volts and **MeV** as a million electron volts.

The mass and energy units that have just been defined are related according to the well-known equation:

$$E = mc^2$$

This emerged from Einstein's theory of special relativity and has the following meaning:

Mass and energy are two manifestations of the same thing, they are as interchangeable as US dollars and Euros in a cash sense, with the conversion factor—the square of c, the velocity of light in a vacuum. Strictly speaking, c is the velocity of electromagnetic radiation, gamma rays, X-rays, visible light, and radio waves, but here it will often be referred to simply as the velocity of light. We will discuss the origins of this equation, and relativity, in more detail, in Chapter 4.

2.5 Photons

Having mentioned the two main players in nuclear physics, the proton, and the neutron, let us now introduce the photon. Electromagnetic energy in the form of light, X-rays, gamma rays, microwaves, and radio waves all travel as small bundles of energy, quanta, The name quanta was introduced by Max Planck in 1900 when he proposed that the different wavelengths in the spectrum of energy emitted by a radiating body

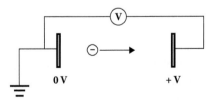

Figure 2.3 The definition of the electron volt as a unit of energy. The electron has V electron volts of energy after being accelerated across an electrical potential of V volts

had to be emitted as small parcels of energy. It was the beginning of quantum theory and became confirmed and generally accepted when Einstein published a paper, using quantum theory, to explain the photoelectric effect. Einstein was eventually awarded a Nobel Prize for this work—not for relativity! Photons are small packets of energy, quanta, that have some similarity to particles, but they have zero mass. However, they carry momentum and can exert pressure, which is strange indeed, and even more so when it turns out they also behave like waves. The nature of light puzzled scientists for centuries. Early experiments showed, beyond doubt, that light was a wave, because interference patterns fitted the theory that light was a wave to perfection. Two waves in step, with their crests coincidental, would give a double size crest, bright light, but if they were displaced so that a crest coincided with a trough, then they would cancel out and the result would be darkness.

Figure 2.4. shows how this would look in the case of Newton's rings. A part of the incident ray is reflected from the bottom curved surface of the lens. Another part is reflected from the flat surface of the glass plate. When these two reflected rays combine

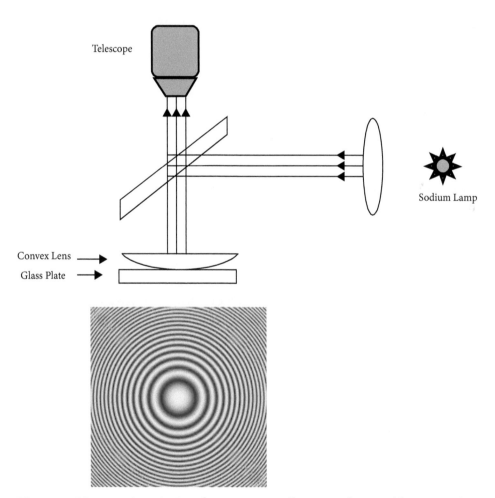

Figure 2.4 Newtons rings. An interference pattern, illustrating the wave-like nature of light

there is a phase difference between them which results in interference. This phase difference changes as we move laterally away from the point where the lens meets the plate. The observed result is a set of light and dark rings, Newton's rings. Clear evidence that light is a wave. So, no problem at all, light is surely a wave! But, if so, how does light get from the sun to Earth through the vacuum, because we know that a wave requires an elastic medium so that vibrations can transmit the wave. Physicists suggested that there must be a substance with elastic properties between the sun and the Earth. Let us give it a name—the ether.

The energy within a photon is related to the frequency of the wave, f, by $E = hf$ where h is a constant, Planck's constant. You can also relate the energy to the wavelength using the relationship between speed, frequency, and wavelength, $c = f\lambda$, where λ is the wavelength; leading to $E = hc/\lambda$.

2.6 Antimatter, Pair Production, and Annihilation

I will now introduce antimatter, which is not a major player in this story of nuclear physics and nuclear reactors, but it does crop up from time to time since positrons have a role in radioactive decay. All particles in the universe have an antiparticle. To give an example, the negative electron's antiparticle is the positively charged version called the **positron**, predicted in the theory of the electron by Paul Dirac and eventually seen in cosmic rays by Carl Anderson in 1932. When matter and antimatter come together, they annihilate each other. When a positron eventually encounters an electron annihilation takes place, turning their mass into energy in the form of photons.

We can now look at two examples of Einstein's equation $E = mc^2$ showing the conversion of mass into energy and vice versa.

The mass of an electron is 5.485×10^{-4} amu which, using $E = mc^2$, is equivalent to 0.511 MeV of energy.

If a photon of energy greater than 2×0.511 MeV enters matter, it has sufficient energy to create two electrons, but to conserve charge it creates an electron-positron pair (see Figure 2.5a), with the electron and positron going off in opposite directions to balance momentum. This process is called pair production and is quite a likely outcome when high-energy photons enter matter. The positron will eventually encounter an electron, and since it is the anti-particle of the electron, they will annihilate each other,

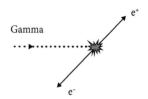

Figure 2.5a Pair production

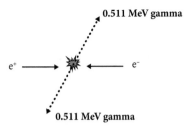

Figure 2.5b Electron-positron annihilation

producing two photons each of energy 0.511 MeV (see Figure 2.5b). This process is called **annihilation**. In the first case energy is changed into mass, and in the second case mass is converted into energy.

2.7 Mass Defects, Q Values, and Cross-sections

The nomenclature for isotopes is to use the chemical symbol, O for oxygen, to tell us the proton number, 8, and the superscript to tell us the total number of nucleons, i.e., protons plus neutrons.

The oxygen in the air we breathe is composed of three isotopes.

99.759% O^{16}, 0.037% O^{17}, and 0.204% O^{18}, all of these are stable isotopes.

In the water we drink, H_2O, the hydrogen is 99.985% H^1 and 0.015% H^2. Both are stable and H^1 is, of course, the proton. We have given a special name to H^2, the deuteron, D, and when D is linked up with an electron to form an atom we call it deuterium. If we have water enriched to a high percentage of D we call it heavy water, often written as D_2O. The nucleus H^3 is not natural but can be made in a reactor, or accelerator, and is a radioactive isotope of hydrogen. It has a half-life of 12.3 years and a special name, triton, with the name tritium being used for the atom. The nucleus He^4 also has a special name, the alpha particle, α.

The radioactive decay of U^{238} by the emission of an alpha particle can be represented by an equation:

$$U^{238} = Th^{234} + He^4$$

The mass numbers, as integers, balance on each side of the equation but the actual mass, in amu, differs from the integer value by an amount called the **mass excess**. In many cases, the excess is actually a deficit and a negative quantity. The mass excess can be converted to its energy equivalent using $E = mc^2$ and is shown underneath each nucleus.

	U^{238}	=	Th^{234}	+	He^4	Q = 4,270 keV
Mass Excess (keV)	47,308		40,613		2,425	

The mass excess amounts on the two sides of the equation don't balance and the difference, 4,270 keV in this example, is called the Q value. A positive Q value indicates

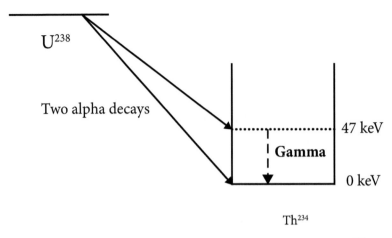

Figure 2.6 An energy level diagram depicting the alpha decay of U^{238}

how much energy is released in the reaction, and a negative Q value indicates how much energy needs to be introduced to make the reaction happen.

An alternative representation of the alpha decay of U^{238} is shown in Figure 2.6. This gives more detail, showing two possible alpha decay branches and the emission of a low energy gamma, 47 keV, when the lower energy alpha is emitted. The Q value of 4,270 keV is shared between the α particle, 4,198 keV, and the Th^{234} recoil, 72 keV.

2.8 Cross-sections

When a neutron approaches U^{235} several reactions are possible. Fission is one of them, and a specific example, where three neutrons are emitted, is shown in the following equation: the Q value is 177 MeV.

$$U^{235} + n = Ba^{144} + Kr^{89} + 3n; \quad Q = 177\,\text{MeV}$$

The probability of this reaction taking place is characterized by a quantity known as the reaction cross-section, denoted by σ (sigma). Each nucleus can be considered as a sphere with an 'effective' cross-section area, σ cm^2, representing the probability of a particular reaction. This is not a real area, it is a conceptual area representing the probability that a reaction will take place, and it is used to calculate the rate of a particular reaction. Different reactions may be possible, each one having its own cross-section.

To illustrate the use of this 'effective area' consider a slab of material with an area A and thickness dx, see Figure 2.7. Suppose it contains N nuclei per cm^3. The total effective (cross-sectional) area of the nuclei in the slab is N.A.dx.σ, which means the probability of a reaction is (N.A.dx.σ)/A, the area cancels out, leaving N.dx.σ. for the probability of a reaction.

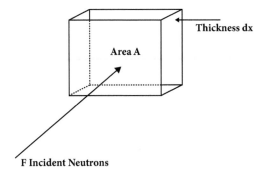

Figure 2.7 Neutrons incident on a slab geometry

If we consider an incident neutron beam intensity of F, then the number interacting will be F.N.dx.σ. If dF is the small change in F due to the reaction, then:

$$dF = -F.N.dx.\sigma, \text{ the minus sign indicating a decrease in F}$$

This is recognizable as an exponential decrease as the thickness increases.

In nuclear physics the unit for this effective area, σ, is called a **barn**, symbol **b**, where 1 barn = 10^{-24} cm^2. The name originates historically from the feeling that such a cross-section was as large as a barn.

There are several versions of cross-section—microscopic cross-section, partial cross-section, total cross-section, macroscopic cross-section, and mean free path, all related to the same 'effective area' concept. The microscopic and partial cross-section are the same and are related to a very specific and limited outcome of a reaction. The fission cross-section in the previous example is a **partial cross-section** since it relates to a very particular outcome of three neutrons and two specific fission fragments. It is much smaller than the total cross-section for fission when all fission channels have been included. The most common use of the term **total cross-section** is the sum of the cross-sections for all possible reactions.

The macroscopic cross-section Σ, is defined as Nσ, the total effective cross-section for all the target nucleons per unit volume. It is seldom quoted itself but its reciprocal 1/Σ can be shown to be the average distance of travel before an interaction. This is called the **mean free path,** which is helpful in appreciating the significance of a reaction in real circumstances. Consider thermal neutrons incident on a target of natural uranium metal. The density is 19.1 g/cc and the U^{235} isotopic content is 0.72%. N is given by:

$$N = 19.1 \times \text{(Avogadro's number)} \times .0072 / 238, \text{ giving } N = 3.47 \times 10^{20}$$

Assuming a fission cross-section of 571 barns gives a macroscopic cross-section as 0.198 cm^{-1} and a mean free path of 5.05 cm. This tells us the average distance travelled between fission reactions is 5.05 cm.

A neutron can get into a nucleus without difficulty since it has no Coulomb barrier to overcome, and the neutron cross-section, when thermal neutrons are captured in

Xe^{135}, is as large as 2.65×10^6 barns. This cross-section, over 2 million barns, is the largest neutron cross-section in nature. Charged particle reactions, on the other hand, are restricted by the Coulomb barrier and are often below a barn,

2.9 The Discovery of Radioactivity

The word nuclear was unknown as a branch of physics until the end of the nineteenth century. In the 1890s scientists were about to realize that there was more to the world than classical physics. The discovery of X-rays in 1895 took the lead, and the theoretical proposal that energy existed in small packages called quanta followed. Einstein's equation $E = mc^2$ appeared soon thereafter, in 1903. Pitchblende, the uranium ore, had become available for study in Europe and understanding the nature of the radiations emanating from this material would be our entrée into nuclear physics, presenting a challenge to scientists for many years.

The door into the world of nuclear physics was first unlocked by Henri Becquerel in 1896. He noticed that radiations from uranium were affecting photographic plates, very similar to Wilhelm Rontgen's discovery of X-rays the previous year. It was a new and puzzling experimental observation. The usual story, that the rays were readily identified by their deflection in a magnetic field, alphas bend to the right, betas to the left, and gammas not at all, is fiction, serving only to reflect the knowledge much later, in 1910. It's an interesting story, a real mystery thriller, which would continue for 37 years until Fermi closed the book with his theory of beta decay, brought about by his suggestion of a new fundamental force in nature.

The experimental problem was severe because there were three different types of radiation, and the uranium samples contained many different decay-chain products to varying degrees. Figure 2.8 shows the decay chain of U^{238} illustrating the complexity of samples derived from the ore, pitchblend, without chemical separation procedures. Measurements from samples using different separation processes would often give conflicting results. Moreover, the instruments of that era were not yet sophisticated enough to make reliable quantitative measurements.

On the theoretical side, asking the question what is happening suffered somewhat because X- rays had just been discovered and it is not surprising that initial thoughts were in the direction that the effect on the photographic plate was due to a form of light. In France, where the action was taking place, the fashionable topics in physics were luminescence, phosphorescence, and fluorescence, the process by which one form of light could change into another. The consequence was that Becquerel's first guess was that uranium rays were a form of light that was convertible into X-rays. To think that particles were being emitted spontaneously would have been an imagination-leap too far.

Maria Sklodowska was a Polish physicist of exceptional ability. She went to Paris in 1891 and decided to investigate 'Becquerel's Rays' for her PhD studies at the Sorbonne. She married Pierre Curie, and the name Marie Curie came into being. She became a

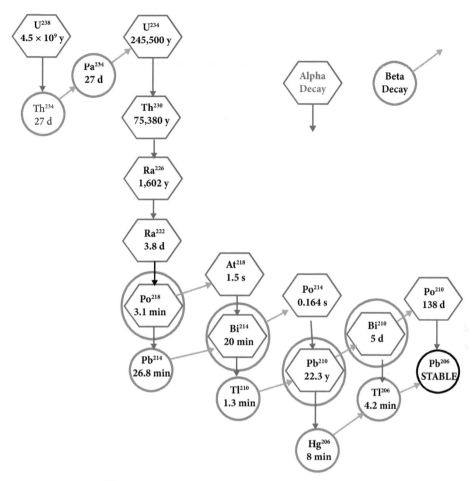

Figure 2.8 The U^{238} decay chain

nuclear chemist as she searched for other heavy nuclei that emitted similar rays. She soon found polonium and radium but the mystery and understanding of radioactivity continued to perplex the scientific community. It eventually dawned on investigators that some of the rays were particulate in nature. A breakthrough came when Becquerel himself identified electrons in the radiation as being the same charge to mass ratio as cathode rays, electrons, so not a form of light after all. There were rumours that atoms might be transforming from one type to another, but it was a dangerous suggestion since alchemy was a no-go topic that could ruin the reputation of any scientist. Chemical separation of the various decay products contaminating the uranium helped to remove some components of the decay-chain and did lead to new information on the nature of radioactivity. In 1900, Rutherford, working in Canada on thorium emanations, discovered an activity that decayed with an exponential decay and half-life of about 1 minute. It was now apparent that uranium ores contained many different sources of radioactivity. Radium became commercially available and could be bought in a shop for eight shillings per milligram. A sizeable source of radium generated its own heat and could bring a similar mass of water to the boil, not just once but time and time again! The mystery deepened and it now became acceptable to talk about nuclear transformations, with an unexpected new source of energy somewhere inside an atom. Bear in mind that the solar system model of the atom was still years away.

The final piece in the jig saw was the identification of the alpha particle as a He^4 nucleus. The mass was finally confirmed by Rutherford at Manchester in 1908 with the help of the recently developed Geiger Counter.

2.10 The General Characteristics of Radioactivity

Radioactivity was now confirmed as having three components: alpha, beta, and gamma. The two main characteristics of radioactivity are that first of all, it is a random process where each atom, or nucleus if you prefer, has a very low, but not quite zero, probability of decaying, and it is impossible to say when any particular nucleus will decay. Observing a decay is only possible because this low probability is counterbalanced by the enormous number of nuclei per gram of material. The second characteristic is the rather obvious fact that if you double the mass, you will double the rate of decay. This leads to the well-known exponential decay law.

The mathematical expression for exponential decay is,

$$N(t) = N(0)e^{-\lambda t}$$

saying that the rate of decay after a time t, N(t), is related to the rate at time t = 0, N(0), multiplied by the exponential function, $e^{-\lambda t}$. The decay constant λ is a characteristic of the material and $1/\lambda$ is equal to the mean lifetime for decay. A more convenient parameter for specifying the speed of decay is the **half-life**, the time taken for the activity to drop by 50%.

This is often written as $T_{1/2}$ and is related to λ by:

$T_{1/2} = \lambda/0.693$; the number 0.693 is the natural logarithm of 2, ln2 = 0.693

The units of radioactivity were originally expressed in Curies, corresponding to 3.7×10^{10} disintegrations per sec, but the modern standard is the **becquerel**, simply the amount of material that emits one disintegration per second.

Another concept, used when speaking about radioactivity, is the **specific activity**. This is the activity per kilo of material. For a nucleus with a long half-life, like uranium, the specific activity will be low, but if the nucleus has a short half-life the specific activity will be high.

Radioactive decay creates a different element, which itself may be radioactive, a daughter product, and this, in turn, may decay further until eventually the decay chain ends with a stable nucleus. Nobel prizes were to be won in this rich field of new discoveries and Marie Curie was honoured in 1903, jointly with her husband, for her pioneering work on radioactivity. She became the only woman to get two Nobel prizes with her second award for Chemistry in 1911, citing her discovery of polonium and radium.

2.11 Gamma Decay

The gamma rays observed by Becquerel were not due to a completely different mode of decay, they were incidental gammas, all produced as a consequence of α or β decay, as the daughter nuclei decayed to the ground state.

2.12 Spontaneous Fission

There are a few other types of radioactivity, beyond alpha, beta, and gamma. Some nuclei decay by neutron emission and since these are very relevant to nuclear reactors, where they are referred to as delayed neutrons from fission, they will be explained later. Fission can also take place spontaneously, without needing a neutron to initiate the process. Several nuclei have a small probability of undergoing spontaneous fission, U^{238} and Pu^{240} are just two examples, but one, artificially produced nucleus, californium Cf^{252}, is a prolific source of neutrons, an enormous 2.4×10^{12}/g, and all from spontaneous fission. It is often used to aid the start-up of reactors and to test and calibrate neutron detectors.

This chapter has covered all the basic physics that is required to understand the fission process and nuclear reactors. To complete the picture and give a deeper understanding of the physics, I explain some of the basic concepts of quantum theory in the next chapter. You may be confused by the different names that are given to the physics of quantum theory. In a fundamental sense, quantum theory is only the theory of the consequences of the fact that Planck suggested that energy is not continuous, it exists in bundles, or packages, that he called quanta. The composite terms quantum mechanics and wave mechanics are used to deal with the fact that particles have wave-like properties, and we need a physics theory that must be capable of dealing with both the particle aspects, mechanics, and the wave aspects, simultaneously. The fact that energy is quantized is a separate requirement but still applies. Wave mechanics and quantum mechanics are often simply expressed by the term quantum theory.

3

Basic Quantum Theory

3.1 Skip This Chapter If You Wish

Some of the concepts that emerged during the development of quantum theory in the 1920s are directly relevant to the detailed understanding of nuclear physics. A notable example is the theory of alpha decay, which relies on Heisenberg's uncertainty principle. However, spending time on the content of this chapter is not required for the understanding of how reactors work. You can skip basic quantum theory and move on to the next chapter.

3.2 The Uncertainty Principle

The dual, particle/wave nature of the photon, was beginning to be accepted at the beginning of the twentieth century. It was a very satisfying surprise in 1924 to read about the theoretical prediction that particles, such as the electron and proton, also have wave-like properties. The theoretical prediction that all particles have an associated wavelength came in 1924 from a hypothesis in Louis De Broglie's PhD thesis. Waves can behave like particles, as is seen in the case of the photon, and particles can behave like waves. He proposed the relationship,

A particle has a wavelength λ given by $\lambda = h/mv$,
This is referred to as the De Broglie wavelength of the particle.
Where h is Planck's constant, m the mass, and v the velocity of the particle.

The wave-like nature of the electron was soon confirmed experimentally, in 1927, when Davisson and Germer showed that electrons could produce a diffraction pattern. Wow, an electron is both a particle and a wave!

So, if we think of an electron as a wave, what is its position? We can no longer think about it as a ball, with a centre of gravity as a point, with no size, defining its position. Heisenberg's uncertainty principle recognizes the fact that the position of a 'wave' is not precise and suggested we should consider the wavelength associated with a particle, λ, as representing the uncertainty in its position. He used the De Broglie wavelength to arrive at his first uncertainty relationship:

$\lambda = h/mv$ can be re-arranged as $\lambda.\ mv = h$

Using 'p' to represent mv, the momentum, it becomes $\lambda.\ p = h$

Understanding Nuclear Reactors. Brian Hooton, Oxford University Press. © Brian Hooton (2024).
DOI: 10.1093/oso/9780198902652.003.0003

since we now accept that λ represents the uncertainty in position Δx, we can replace λ with Δx, and p, the momentum, can be replaced by Δp, the uncertainty in momentum. This results in:

$$\Delta x \, . \, \Delta p = h \text{ as a statement of the uncertainty principle}$$

In words, the uncertainty in position multiplied by the uncertainty in momentum is equal to Planck's constant.

However, the equal sign should be taken with a pinch of salt because the word 'uncertainty' has not been defined. A more precise equation came later when uncertainty was linked to 'standard deviation'.

The uncertainty relation tells us that, if the uncertainty in position is small, then the uncertainty in momentum is large and vice versa.

There is an alternative version of the uncertainty principle, which can be derived using the fact that mass and energy are interchangeable. It is,

$$\Delta E \, . \, \Delta t = h \text{ relating uncertainties in energy and time.}$$

This can be interpreted in several ways. It allows the conservation of energy to be violated temporarily, by saying that since energy is uncertain, energy can be borrowed, provided the amount borrowed, ΔE, is repaid within a time Δt, where $\Delta E \, . \, \Delta t = h$

ΔE also relates the width of a bound energy level in a nucleus to its lifetime. If a nuclear level has a broad resonance, it is not a definite energy, it has a large energy uncertainty and will therefore have a short lifetime for decay. Conversely, if the energy width is small, it will have a long lifetime. If energy can be borrowed, temporarily, then mass can also be created, temporarily. This possibility to create mass is linked to the creation of bosons, which have mass but may only exist for a short time.

3.3 The Theoretical Treatment of Nuclear Physics

In a nuclear reaction that leaves a nucleus in an excited state there are several possibilities for what happens next. The excited state could decay by gamma ray emission either direct to the ground state or by a cascade producing several gammas. It might also be able to decay by emitting a particle. All these options are referred to as exit channels. The various exit channels available from an excited state posed a fundamental problem. Surely the laws of physics dictate that once the excited state has been formed the next step is predicted, and unique. How can it be possible that if we populate the state a second time it can do something different, since the next step, according to classical physics, should be determined. This philosophy was called **determinism** and many notable physicists, including Einstein, supported it. He famously said God doesn't throw dice! Moving away from determinism was the challenge to all the theoretical physicists in the 1920s. From a theoretical standpoint how could you describe this excited state if there were many options available for what happens next. The answer was to define what happens next in terms of probabilities. A theory was needed that could describe the excited state as a superposition of all the possible exit channels. In the 1920s theoreticians of that period, Heisenberg, Schrodinger, Fermi, Dirac, and many others did manage to produce versions of quantum theory, based on probabilistic outcomes. Their theoretical approaches, wave mechanics and quantum mechanics, were different in their mathematical formulation, but the results were more or less equivalent, and in agreement with each other. Eventually the theory caught up with experimental observations and actually started to move ahead, predicting what should be observed in the future. In the second part of the twentieth century, the theoreticians were able to predict new particles, such as the Higgs boson, way before the experimentalists were able to confirm their existence.

One of the main concepts of wave mechanics is the wave function, which describes a particle or an energy state. The wave function for an electron can be considered as giving it a probability of being inside or overlapping the nucleus to some extent. The electron is no longer a point in space. It is, so to speak, smeared out. This means that the nucleus and the atom should not be considered as completely separate, they are a combined quantum system. There are several examples of this overlap.

If a nucleus is in an excited state and has a probability of decaying to its ground state by emission of a gamma ray, quite a common occurrence, it can also decay by ejecting an electron from the K shell, or L shell of the atom. This mode of nuclear decay demonstrates that the nucleus can feel the presence of the electron. It is called **internal conversion.**

Another process where the nucleus and the atom act in partnership is when the nucleus captures an electron from the K shell, called K capture. This can only happen when the nucleus is radioactive and capable of emitting a positron. The nucleus losing a positive charge via a positron is equivalent to it gaining a negative charge via the K capture.

The nucleus is not like the atom, it does not have a central core with nucleons moving around in orbits. The forces between nucleons are short range strong forces of attraction and it is a mistake to try and picture what is going on in terms of nuclei orbiting around a central core because there is no accurate picture. The only way to understand the internal workings of the nucleus is by theoretical equations and using a theoretical representation of the nuclear forces operating inside the nucleus leads to a shell model, with several added variations.

It soon became apparent that, inside the nucleus, a theory of shells, similar to the atomic theory, would work. The energy levels could be predicted and transitions between energy levels, releasing gamma rays, would be limited by selection rules. Energy levels in the nucleus are defined by quantum numbers and shells become full according to the exclusion principle, only one particle in one state. The nuclear shell model was only a partial solution since the nucleons also have the ability to have rotational motion and even vibrational motion, superimposed on the shells. When everything is put together the nuclear model is called the unified model. The energy levels are specified by their spin and parity (see section 3.7), which participate in the main selection rules for transitions. Other properties, such as the shape of the nucleus (not a sphere), magnetic moment, and rigidity (moment of inertia), also feature in our complete theoretical understanding of the interior of the nucleus.

3.4 Atomic Spectra and Quantum Numbers

As far back as 1756 scientists had been recording the coloured light emitted by elements that were hot. Thomas Melville did it by adding salts to alcohol flames and by 1853 the Swedish scientist, Anders Angstrom, was producing spectra from gases. Patterns in the line structure were found and given names, The Balmer Series, the Lyman, and the Paschen series were identified in the nineteenth century. Johannes Rydberg, another Swedish scientist, did find a formula that would explain some of the series but understanding these patterns was a difficult task with a solution that should lead to the classification of groups of elements such as the inert gases, the halogens, and the alkali metals in the Periodic Table. The task of explaining the patterns was beyond the capability of the scientists working in the field of atomic spectra. It would remain so until Niels Bohr defined the internal workings of the atom as a solar system in 1913. He suggested an atom had electrons moving like planets, in orbits around the nucleus. This opened the door to a much better understanding of the ultra-microscopic world. Good-quality and accurate data were now available on the light emitted when elements were excited in an electrical discharge. These data included spectra taken in the presence of a magnetic field (Zeeman effect) and an electric field (Stark effect), both revealing additional lines in the spectrum. Now was the time to seek an explanation. Atomic spectrum lines provided a wealth of information on energy levels in atoms; a twentieth-century example is shown in Figure 3.1. They demonstrated that

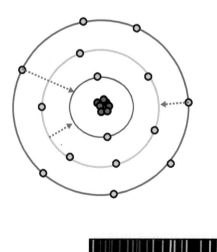

Figure 3.1 Showing discrete line spectra when electrons fall into lower energy levels

the energy levels were discrete, with a fixed energy value, and that transitions between states would result in the emission of a photon at a particular energy, a spectral line. It should not be difficult to discover the set of rules governing the observed spectra, and the physics required to explain them.

3.5 Sommerfeld's Contribution

Arnold Sommerfeld was a veteran German theoretical physicist, a year older than Einstein. He would become famous, not only for his work on atomic spectra, but as the supervisor of numerous eminent physicists of his day. These included Wolfgang Pauli, Werner Heisenberg, Peter Debye, and Hanse Bethe, all Nobel prize winners.

Sommerfeld started to think about the rules necessary to explain the spectrum lines and also to explain the grouping of elements in the Periodic Table. The reason for the occurrence of each inert gas, helium, neon, and argon, with electron numbers 2, 10, and 18, was a particular challenge. Sommerfeld plucked rules out of thin air to see if they would fit and, after some limited success, he searched for reasons to explain the rules. He identified rules based on three quantum numbers, a main electron orbit, and two sub orbits. These were given by letters, n to describe the main orbit and k and l to describe the sub orbits. These three numbers could be considered as specifying the address of each electron. An electron's address, with just three quantum numbers, is equivalent to saying they live in a building number, with a floor number and a room number. When the building is full you have an explanation for the inert gases. These three quantum numbers were also capable of explaining all the complexities of atomic spectra, apart from one very significant anomaly!

3.6 Pauli's Contribution

Wolfgang Pauli was an Austrian physicist born in 1900. He was a teenage wonder and a protégé of Einstein, who obtained his PhD at the age of 21. It was no surprise that he had an aptitude for physics since his godfather was Ernst Mach, the originator of Mach number 1 as the speed of sound. His PhD supervisor was Arnold Sommerfeld. When Pauli was seen wandering about Copenhagen with a frown on his face and challenged, he responded by saying, 'How can one look happy when he is thinking about the anomalous Zeeman effect', which is the appearance of strange spectral lines in the presence of a magnetic field. Sommerfeld himself had toyed with the idea of a possible fourth quantum number but it was very much an ad hoc thought. Pauli now resolved everything with a more comprehensive solution to the anomaly. He suggested that yes, there needs to be a fourth quantum number giving the electron a two-valued property. This would provide the extra dimension and we also need a new principle saying: the quantum numbers identify a unique particular state which can only be occupied by a single electron. The word he used 'occupied' can be identified as an address. This is the

Pauli Exclusion Principle. There was no suggestion, at this time, as to what the two-valued property of the electron, the fourth quantum number, related to: no mention of spin.

3.7 Spin and Parity

The exclusion principle was suggested by Pauli in 1924, two years after Otto Stern and Walther Gerlach had carried out their ingenious experiments in 1922. They had produced a beam of silver atoms from a hot oven and had discovered that if the beam passed through a magnetic field, it split into two. Their explanation for this was that the electron seems to have a magnetic moment and behaves like a small bar magnet. This idea looked promising at first since a rotating charge would exhibit a magnetic moment. However, the calculations indicated that the electron would have to be spinning at a speed greater than the velocity of light. Nobody was ready to accept their explanation.

The next suggestion on the two-valued property of the electron came from an American physicist Ralph de Laer Kronig. He suggested that the two-valued property was associated with the self-rotation of the electron; the word spin comes to mind! He was a respected physicist who made calculations to examine the effects of his suggestion and found differences between measurements and his predictions. He was connected well enough to discuss it with Pauli, Bohr, and Heisenberg, who more or less poured cold water on his suggestion, so he let it drop.

Less than a year later, two Dutch physicists, Samuel Goudsmit and George Uhlenbeck, independently suggested the same thing and sent a paper off for publication. They spoke to the eminent physicist Lorenz who said it was rubbish and impossible in classical electron theory, so they tried to withdraw the paper, but were too late. It was published in November 1925. This time the physics fraternity in Copenhagen paid attention. An assumption that the electron could entertain two, and only two, states of angular momentum which could be coupled to the orbital angular momentum answered all the unanswered questions. The two states had to be an angular momentum of $+\frac{1}{2}$ or $-\frac{1}{2}$ but the word spin was not yet used, it was referred to as self-rotation. After a while the word spin started to be used, even though there was universal agreement that it was not the same as a small spinning top, the classical understanding of spin didn't fit. Perhaps surprisingly, the term 'spin' exists today; what else could you call it!

Parity is not so easy to explain. It is important because it is an entity that is considered to be conserved. If a suggestion by a theoretical physicist did not conserve parity it would be ignored. Energy and momentum are conserved, and we can identify them with aspects of physics that we are familiar with. Parity is a mathematical property of the wavefunctions that describe matter in the world of quantum mechanics. It relates to reversing the coordinates, x, y, and z, in a theoretical description of the energy state, which means reversing the coordinates of the wave function. This is often referred to as looking at the mirror image. If the mirror image is the same, we have positive parity. If it is different, we have negative parity. The letters A and H are the same in a mirror image, so can be said to have even parity The letters B and R have negative parity.

The quantum mechanical rules state that parity and spin are both conserved. However, it should be noted that in 1956 Lee and Yang were awarded the Nobel prize for identifying a particular case where parity is not conserved. This was in beta decay, which is explained in Chapter 3 as due to the weak nuclear force.

3.8 Alpha Decay

When Rutherford identified the alpha particle as He^4 in 1910 there was only one difficulty. It was quite acceptable that they existed inside the nucleus but how did they get out? The nucleons would share energy and try to escape all the time, but they did not have enough energy to break the strong nuclear force holding them together in the nuclear potential well. They could not climb over the Coulomb barrier. George Gamow, a Russian physicist, suggested an explanation in 1928. The process is illustrated in Figure 3.2. The alpha can borrow energy, within the rules of the uncertainty principle. It can use this borrowed energy ΔE to get over the barrier provided it pays back the debt within a time Δt according to the Heisenberg Uncertainty relationship $\Delta E.\Delta t = h$. The process was referred to as **tunnelling**. Since tunnelling, with the benefit of the uncertainty principle, is a very unlikely event the half-life for alpha decay is, in most cases, very long and uranium has survived in the Earth's crust to outlive the age of our planet. U^{238} has a half-life of 4.468×10^9 y.

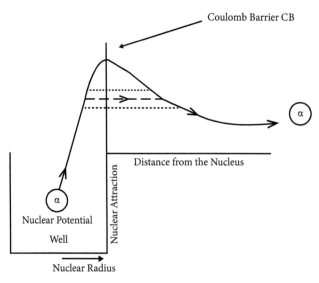

Figure 3.2 The path of an alpha particle escaping from the strong nuclear forces in the potential well by tunnelling through the Coulomb barrier

3.9 Beta Decay and the Story of the Neutrino

The process of beta decay is the emission of an electron or a positron. Although the electron was soon identified by Becquerel himself in 1900, it was not until 1934 that the positron decay of P^{30} was recognized. The positron is the antiparticle of the electron, but antimatter was not suggested until the late 1920s.

Paul Dirac was an enigmatic theoretical physics who was the Lucasian Professor of Mathematics at Cambridge, following in the footsteps of Isaac Newton. Einstein wrote, 'I have trouble with Dirac. This balancing on the dizzying path between genius and madness is awful'. Dirac's contemporaries at Cambridge defined a new unit of speech, the 'Dirac', meaning one word per hour! His brilliance became evident with his theory of the electron in 1928, which, when subjected to its consistency with the theory of relativity, actually predicted, or suggested, the existence of antimatter. Dirac himself did not suggest antimatter but anti-electrons were soon identified in cosmic rays by Carl Anderson, and the antiparticle to the electron was named the positron. In radioactivity, if positron decay is possible, then electron capture, the nucleus capturing an electron from the K shell of the atom, is always possible since the endpoint is the same. Lose a positive charge or gain a negative charge.

Understanding beta decay perplexed physicists for more than 30 years. The first problem of many was—where are these electrons coming from? They are not present in the nucleus, or are they? Is it possible that N^{14} could be 14 protons and 7 electrons? This suggestion was not completely discounted until James Chadwick discovered the neutron in 1932. Chadwick had made precise measurements, in 1914, confirming that the energy of electrons in beta decay varied, and had a spectrum running from zero energy up to a specific value, E_{max}. This corresponded to the Q value and would account for all the energy available from the mass defects. The obvious question is—when the electron energy is less than E_{max}, where is the missing energy? Is another particle involved? At some stage it was realized that there was also a problem with spin. The emission of an electron changed the element but not the mass number. A nucleus with integer spin should stay integer, and a nucleus with half-integer spin should stay half-integer. This was impossible with the electron removing a spin of ½.

Matters came to a head in 1930 when Pauli was bold enough to suggest, in a letter to Enrico Fermi, that the nucleus contained a very low-mass neutral particle, which he called a neutron. In 1931 Fermi responded by making a counter-suggestion that it should be called a neutrino, a little neutron in Italian. Perhaps he had anticipated Chadwick's discovery of the real neutron in 1932?

Conservation of spin was recognized as a problem since although spins combine as vectors and can be considered as 'up' and 'down', expressed as spins of +1/2 or −1/2, the beta decay $Th^{234} = Pa^{234} + e$ is not possible since the spins cannot balance.

The answer to the missing energy, and the spin, is the third particle, **the neutrino, ν.**

If we add a neutrino to give $Th^{234} = Pa^{234} + e + \nu$ the spins can balance.

It was eventually recognized that the neutron was slightly heavier than the proton and was not in fact a stable elementary particle. It decayed by beta decay, with a half-life of 10.2 min and a Q value of 0.78 MeV, n = p + e + ν. Please note, the neutrino associated with electron decay is actually an anti-neutrino, the reason being conservation of leptons—not part of this narrative!

Fermi provided the solution to the theory of beta decay with his landmark paper in 1933, the final piece in the jig saw. He said that neutrinos and electrons do not exist in the nucleus, but are created at the time of beta decay by means of a new force in nature, the weak nuclear force. It uses the energy available from the mass defect. A new force in nature was indeed a remarkable suggestion but it explained everything. The neutrino was, at the time, expected to have zero mass and no charge, it was nothing, spinning!

Proving that the neutrino existed required experiments that seemed impossible. The weak nuclear force was very short range, and a neutrino would pass right through the Earth with little chance of any reaction. Scientists eventually designed systems that should be capable of detecting neutrinos if the number of incident neutrinos was large enough. Sure enough, this eventually happened when very expensive and successful neutrino detection systems started to crop up in several countries and it was realized that the sun was a prolific source of neutrinos.

The neutrino story continued to intrigue physicists as elementary particle physics progressed and three types of neutrino were identified. They were associated with the electron and two elementary particles, the tau and the muon, they were referred to as neutrinos with three flavours. Interest in the neutrino continued when it was observed that, when neutrinos were travelling through space, one flavour could flip over to another flavour, a process called oscillation. A tau type (flavour) could suddenly become an electron type. For this oscillation to happen, the neutrino would have to have some mass, definitely not zero, but it could be very small. The neutrino seems to have mass, by implication, but we don't have a value.

It is now suggested that a fourth type of neutrino may exist, already named as the sterile neutrino because it would only interact through gravity. This new type of neutrino is linked to a suggestion that it may provide an entrée into the unknown world of dark matter. Dark matter seems to be needed to explain the lack of gravitational forces in the universe—another story for another day.

3.10 The Discovery of the Neutron

The story of the neutron is linked to the question—do electrons exist in the nucleus? The existence of a negatively charged particle in the nucleus posed several problems and several scientists had suggested the existence of the neutron throughout the 1920s. It fell to **James Chadwick** to identify the neutron and receive a Nobel prize for its discovery in 1935.

Chadwick, born in 1891, was a prodigy who intended to study mathematics at Manchester University. During the reception process, he apparently stood in the line for physics by mistake but was too embarrassed to do anything about it when he arrived at the head of the queue, so he signed up for physics instead. He lived at home and walked the 4 miles to and from the university each day. He graduated at the age of 19 and in 1910 went to study with Hans Geiger in Berlin. When the First World War broke out, he was interned. He returned to work at Cambridge with Rutherford and heard about experiments carried out by bombarding beryllium with alpha particles from a radioactive source. The irradiation produced very penetrating radiations which were erroneously said to be gamma rays. Chadwick repeated the irradiation using a small amount of polonium (0.5 µg) sent to him by Lise Meitner from Germany. This was the same Lise who was awarded a Nobel prize for the discovery of fission. The reaction induced by alpha radiation of beryllium we now know as:

$$Be^9 + \alpha = C^{12} + n$$

He confirmed the reaction produced particles, not gammas, by observing protons knocked out of paraffin wax into sensitive proton ionization chambers. He immediately sent a letter to *Nature* titled 'Possible Existence of a Neutron' and then followed it up with a more detailed paper to the Proceedings of the Royal Society titled 'The Existence of a Neutron'.

Fermi's 1933 paper was the culmination of the experimental discovery of radioactivity, which began in 1896, and was the catalyst for our understanding of the nucleus, embracing the new strong and weak nuclear forces and all the understanding revealed by the eminent physicists of the 1920s. From the 1930s onwards, new tools for understanding the detailed inner workings of the nucleus were becoming available, particle accelerators such as the Van de Graff and the Cyclotron were used to bombard all the nuclei in the Periodic Table with protons, alphas, and many other projectiles. New radiation detectors provided the results to give a detailed understanding of nuclear structure that we now possess.

3.11 Quantum Theory and Beyond

We can conclude this chapter with a review of quantum theory, at its stage of evolution when fission was discovered in 1938.

Quantum theory emerged gradually in the 1920s and adopted several new concepts, way beyond the classical era. They were required to explain atomic spectra, radioactivity, the absence of determinism, and the dual particle-wave nature of matter. We can summarize these new concepts as follows.

Max Planck suggested that energy was not continuous but contained in small packages called **quanta.**

Einstein's theory of relativity revealed the fact that mass and energy were two manifestations of the same thing, related by $E = mc^2$.

The optical line spectra, seen when atoms were excited in a discharge, suggested several rules to explain them. The exclusion principle allows only one electron in a specific energy state where the states are identified by quantum numbers, a form of address. The electron possessed a two-fold condition, called spin. Transitions between energy states would result in the emission of a photon, but some transitions were forbidden by selection rules.

Clear evidence emerged that photons and particles could show both particle and wave characteristics and they possessed these simultaneously. This implied that a particle with wave characteristics can only have its position specified to the accuracy of its wavelength. This was set out as the uncertainty principle.

The classical approach of determinism, where arriving at a particular state would determine what happens next, did not apply to the ultra-microscopic world. It had to be replaced by a set of probabilities to account for several possibilities of what happens next.

Quantum mechanics, or quantum theory, is the theoretical way of describing this world, but it suffers from the fact that it lacks visualizability. I suspect this ugly word was invented by theoretical physicists in the 1920s. It is not easy to see an electron in the form of both a particle and a wave at the same time, simultaneously. The best one can do is shown in Figure 3.3. where it is represented as a 'wave-packet', smeared out and not localized. The lack of a visual picture made it controversial and caused many arguments between the eminent Nobel Laureates of the nuclear fraternity in the 1920s. Since it is not visualizable we must rely on a mathematical approach to give us an accurate picture. The first theoretical breakthrough came with Heisenberg's matrix mechanics' approach. This was closely followed by Schrodinger's wave mechanics, both gave similar results but, at the same time, both had shortcomings. Nevertheless, quantum theory had arrived and was capable of dealing with most of the experimental observations. The few exceptions were eventually resolved with Richard Feynman's development of an extension to Maxwell's electromagnetic equation under the heading of quantum electrodynamics (QED).

Quantum theory and QED were not the answer to everything and did not address the nature of the strong and weak nuclear forces. This aspect of mother nature was about to explode with the identification of a host of new elementary particles. I am talking about the world of quarks, fermions, baryons, bosons, leptons, neutrinos, and hadrons, not to mention the God particle! The conclusion to this development, with a reasonable understanding of the numerous, newly identified elementary particles, is referred to as 'the Standard Model'. It is a categorization of all the elementary particles classified into defined sets. Fortunately, we can understand the operation of nuclear reactors without any knowledge of the Standard Model, so we can decline the option to expand our knowledge in this area.

Wave Packet

Figure 3.3 A particle depicted as a travelling wave packet

The main features of quantum theory came from the minds of many physicists. De Broglie's wavelength for a particle is used to explain the Heisenberg Uncertainty principle. Determinism is replaced by wave mechanics based on the probability of decay through exit channels. The concepts of spin and parity are explained with contributions from Pauli, Ralph de Laer Kronig, Samual Goudsmit, and George Uhlenbeck. Sommerfeld uses quantum numbers to explain atomic spectra. Pauli uses observations of the anomalous Zeeman effect to suggest new quantum numbers based on the electron having an additional two-value property, spin. Lee and Yang show parity is not always conserved. The theory of alpha decay using George Gamow's tunnelling explanation and Fermi's explanation of beta decay following Pauli's suggestion of spin and the prediction of the neutrino complete the explanation of radioactivity. Dirac predicts antimatter and positron decay is observed. And James Chadwick discovers the neutron.

4

The Story of E = mc² and Relativity

4.1 The Unification of Electricity and Magnetism

Einstein's equation came as a surprise, even to Einstein, but its meaning is quite clear and not difficult to understand. Energy and mass are interchangeable, but why is c, the velocity of light, the conversion factor, and how did it get involved?

The story of E = mc² is the story of relativity, which begins way back in the year 1799 when Volta invented the battery. This was indeed a remarkable milestone in the march of technology, not only because it would lead to batteries for torches and cell phones, but because it was the very first time that scientists had produced a source of electricity that was stable and reproducible. It gave physicists the ability to study electrical phenomena in the laboratory and to make reproducible measurements. Before the battery the only source of electricity was static electricity, which was fine for producing sparks, but even that depended on the humidity and nature of the frictional surfaces. Volta's battery opened the door to electricity and magnetism and revealed new knowledge that surprised everyone.

In the eighteenth century, it was believed that electricity and magnetism were unrelated. But when Volta informed the Royal Society of his new battery, and demonstrated his invention to Napoleon in Paris, the experimental door to electricity and magnetism was flung open and progress was rapid. New terms in the lexicon of electricity, like Volt, Ohm, and Ampere, soon became essential to define Ohm's law. Within a short space of time, it was observed that an electric current generated a magnetic field and the connection between electricity and magnetism was established. Further observations confirmed that a moving magnetic field would create an electric voltage. Electricity could create magnetism, and magnetism could create electricity. This was the birth of electromagnetism, no longer two separate fields of science.

As experiments progressed in the nineteenth century, with the development of electricity generators and electric motors, the theory was not left behind. James Clerk Maxwell, in 1862, produced a set of equations, Maxwell's equations, that gave a comprehensive theoretical account of all the experimental observations. These coupled equations could be combined and rearranged in a different way and one form was recognized as the standard representation of a wave. It was a wave equation that contained, within it, a component specifying the velocity of the wave. The wave velocity is that of electromagnetic radiation and was theoretically given as a combination of the fundamental strengths of the electric force and the magnetic force. This should make it possible to get a value for the speed of the wave by measuring the strength of

Understanding Nuclear Reactors. Brian Hooton, Oxford University Press. © Brian Hooton (2024).
DOI: 10.1093/oso/9780198902652.003.0004

the electric and magnetic forces in a laboratory. Yes, it is possible to get a value for the speed of light without using a light source. It was realized that this would mean that the velocity of light had a unique and fixed value, and we shall see that this presents a problem! However, several excellent measurements of the speed of light, using a light source, had been taking place in France and the USA using either a rotating mirror technique, developed by Foucault, or a toothed wheel by Fizeau. Eventually it was Michaelson who published a remarkably accurate value of 299,910 +/– 50 km/sec in 1879.

4.2 Relative Motion

The key question was—if this is the speed of light—RELATIVE TO WHAT? We tend to forget that all speeds are relative to something, they must be. There is no such thing as an absolute speed. Two passengers sitting on a train are at rest, relative to each other but the train is moving relative to the track. When the train arrives at a station it is at rest relative to the track but not relative to the sun since planet Earth is orbiting the sun at high speed.

Consider a bullet leaving a gun with a muzzle velocity of V. If the gun is mounted on the wing of a fighter jet, then the velocity of the jet must be added to V to arrive at the bullet velocity relative to the ground.

Physicists in the nineteenth century suggested that the speed of light should be considered as being relative to the ether—that elastic material postulated as necessary for the propagation of a light wave. Michaelson and Morley carried out a very careful experiment to detect the motion of the ether. They measured the speed of light at two different directions at right angles, but the results remained the same, and gave no evidence for the existence of the ether. Enter Einstein, who said, let us assume that the velocity of light is constant and does not depend on the speed of the source emitting the light, totally unacceptable, and contrary to Newtonian physics and to common sense—who could be crazy enough to suggest such a thing, that was lateral thinking in the extreme.

The laws of physics are based on the observation of 'events' which require both a position and a time for definition. A simple event is the kick-off at a football match which takes place at the centre spot in the middle of the pitch **and** at a specific time. Observation of events is what we use to establish the laws of physics but we need to be careful because the speed of light is finite. Consider Figure 4.1. The scene is four piers at the seaside, A, B, C, and D. Tom is at the end of pier B, and Dick is at the end of pier C. Harry is in a boat observing them from a position at sea, equidistant from his two friends. He observes them both to drop off pier B **simultaneously**! However, an observer at A would say Tom dropped off first, because he is closer, and light takes time to travel. An observer at D would say no, Dick dropped off first. So, who is right—was it Tom or Dick who jumped into the sea first?

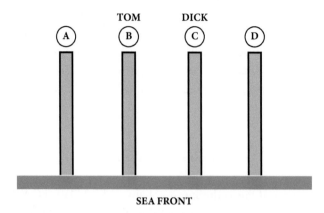

Figure 4.1 What is the meaning of simultaneous?

4.3 Einstein's Theory

Einstein's theory introduced a new definition of simultaneity. He stated that the observations of an event in a coordinate system A must result in the same laws of physics if the same observations were made from another coordinate system, B, moving at a velocity V relative to system A. The two coordinate systems are shown in Figure 4.2. It shows a coordinate system A and a second coordinate system B moving at a constant velocity V in the X direction. Einstein then proposed a set of events in system A and said:

If the velocity of light is constant in both coordinate systems, then—**what would have to be** for the laws of physics in the two systems A and B to be the same? That is:

1. Assume the velocity of light is constant.
2. Assume the laws of physics would be the same if observed from another coordinate system B moving at a speed v relative to A.
3. WHAT WOULD HAVE TO BE?

His first paper on relativity, published in 1903, showed what would have to be:
   Any length in B would have to be replaced by a contracted length.

$$L_B = L_A\{1 - (v/c)^2\}^{1/2}$$

This is referred to as length contraction, with the length tending to zero as the velocity v approaches c.
   Any mass in B would have to be increased according to:

$$M_B = M_A/\{1 - (v/c)^2\}^{1/2}$$

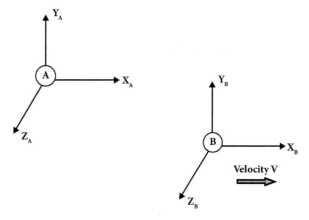

Figure 4.2 Two coordinate systems, B moving at a constant velocity relative to A

The mass would tend towards infinity as v tends towards c.

There would also be a change in time. This is a strange one to understand because we all have a perception of time, either as a point in time or a period of time. We can visualize a length and a mass, but not a time. The time change is referred to as a time dilation according to:

$$T_A = T_B\{1 - (v/c)^2\}^{1/2}$$

These three equations are very close to the classical limit at the speeds we encounter in everyday life. A projectile moving at ten times the speed of sound would still only have a 0.01% change of mass. It is only when we consider electrons and other elementary particles accelerated in machines, produced in nuclear reactions, in cosmic rays and moving at speeds close to the speed of light that we need to correct the values of mass, length, and time.

Einstein had successfully dismissed the ether as unnecessary and must have been delighted with the results. He specified the three equations that must be used to transform the basic dimensions of mass, length, and time. In practice, it means that if high speeds are involved then the relativistic transformations of mass, length, and time must be used. He then looked at further implications of the theory of relativity and discovered, almost by chance, a golden nugget. It was the necessary relationship:

$$E = mc^2$$

This is the well-known Einstein's equation, leading to the atomic bomb and nuclear reactors, with little need to worry about the relativistic transformations of mass, length, and time.

4.4 Standards of Mass, Length, and Time

There are seven units that require definition in the International System of Units, SI. Time, mass, and length are the obvious ones with wide applications but others, the ampere (for electric current), the Kelvin (for temperature), the candela (for luminous intensity), and the mole for the amount of material, need to be defined to cover all the branches of science.

The current time unit, the second, has been well-established since the Atomic Clock was accepted in 1955. The second is defined as exactly 'the duration of 9,192,631,770 periods of the radiation corresponding to the transition between the two hyperfine levels of the ground state of the caesium atom'. This is the best we can do at the moment, and it is quite good enough and accurate enough for most purposes, but it is based on a physical happening which is not constant in an absolute sense.

The units of length and mass were, for many years, based on a platinum bar for the metre and the International Prototype of the Kilogram, a platinum-iridium cylinder; both kept in Paris. The twenty-first century has adopted a different approach to these two standards. The realization that the velocity of light in a vacuum is a fundamental constant gives us the opportunity to define length, the metre, by assigning the absolute value 299,792,458 metres per second as the speed of light. This assignment, adopted in 2019, now defines the metre.

Mass has also been redefined when it was suggested that if we assign a value to Planck's constant, h, we can use it to redefine the kilogram. And so it goes, Planck's constant was given the absolute value $6.626\ 070\ 15 \times 10^{-34}$ in units kg. m^2. s^{-1} to now define the kg. This was announced by the General Conference of Weights and Measures in 2019. It's all very well for scientists to define these units in esoteric terms but the world still needs practical standards and for this purpose, we have turned to the Kibble balance. The Kibble balance is a sophisticated and complex measurement system that has been developed and improved from 1975 to the present day. The technique is now used in all the major standards institutes, worldwide. In 2014 it was used in Canada to provide a value for Planck's constant to an accuracy of nine parts per billion. This opened the door for Planck's constant to be accepted as the basis for a new definition of mass. The Kibble balance can be used to set up precision kilogram masses as standards for everyday use and we should remember Planck's constant every time we buy a kilogram of apples.

5

The Fission Process and the Characteristics of Fission

5.1 The Discovery of Fission

Although there was a good explanation for how an alpha particle could escape from within a nucleus, the strong attraction of the nuclear force made physicists believe that it would be impossible for any nucleus to break up into two large fragments. When fission was confirmed in 1938, and became an established fact, a re-think was necessary. The theoretical answer was the 'liquid drop model', not a new idea since it was first proposed by George Gamow, a Russian theoretical physicist, in 1929. In this model, illustrated in Figure 5.1, the nucleus behaves like a drop of water, with a surface tension preventing the two drops from separating. The neck, in the middle of the drop, weakens the strong nuclear forces holding the drop together and the two parts, each with a large positive electric charge, have sufficient electrical repulsion to overcome the surface tension, enabling the drop to split into two large fragments. The proof of the pudding is in the eating; fission is a reality, it does happen.

The discovery of fission came from a long-standing collaboration between Otto Hahn, a German nuclear chemist, and Lise Meitner, an Austrian physicist. She joined Hahn in Berlin in 1908 and they achieved many successes before the discovery of fission. In the 1930s, they started to study the effect of irradiating uranium with neutrons. The only neutron source they had available at the time was using alpha particles from a radioactive source to bombard Be^9, taking advantage of the $Be^9 + \alpha = C^{12} + n$ reaction as a source of neutrons.

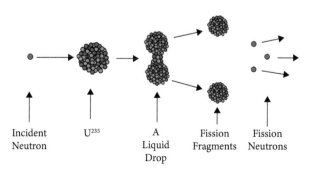

| Incident Neutron | U^{235} | A Liquid Drop | Fission Fragments | Fission Neutrons |

Figure 5.1 The liquid drop model of fission. After the nucleus absorbs a neutron, it distorts into the shape of a liquid drop. Fission takes place with a release of energy and neutrons

Understanding Nuclear Reactors. Brian Hooton, Oxford University Press. © Brian Hooton (2024).
DOI: 10.1093/oso/9780198902652.003.0005

Hahn and Meitner found signs that bombarding uranium with neutrons created barium which, according to conventional theory at the time, seemed impossible. However, distinguishing between barium and radium by chemical analysis was a notorious problem, but it was not a new problem—it had been tackled by Lise and Otto on numerous occasions. They had been working together since 1908 and they were very experienced. They had confidence in their chemistry and were quite sure that it was indeed barium that was being produced, but the nuclear theory said it was impossible. The latest results were being digested in 1938 when the Nazi persecution of Jews became much more evident. Lise Meitner was from a Jewish family but had converted to Christianity as a teenager. This clearly offered no protection, so she left Germany in July 1938 and went to Sweden, via Copenhagen, with the help of the Danish Nobel Laureate, Niels Bohr. In the meantime, Otto Hahn continued his work with a new assistant, Fritz Strassman, and together they confirmed beyond any reasonable doubt that it was indeed barium, not radium. Hahn wrote to Lise Meitner in Sweden who, by good fortune, had arranged to spend Xmas skiing with her nephew Otto Frisch. Frisch had become a notable nuclear physicist and had already spent five years in Copenhagen working with Neils Bohr. Lise and Otto Frisch read the latest letter from Germany and Frisch tentatively suggested that Hahn might have made a mistake. Lise could not accept that and replied by saying that Hahn was too good a chemist and it was not a mistake! Their discussion continued and after a while, considering the forces and the energy balance, the penny dropped: uranium was indeed breaking up into two large fragments and it could, after all, be explained by the liquid drop model. Frisch dashed back to Copenhagen to discuss it with Bohr, who agreed. Frisch then called into the local biology laboratory and coined the term used in biology for the splitting of basic live cells, **fission**. It was his suggested name for the new discovery. Why did he dash back to Copenhagen?

5.2 Niels Bohr and Copenhagen

Bohr's insight into the inner workings of the atom had put him on a pedestal almost as high as Einstein's. He was revered in Denmark, Scandinavia, and the rest of the world, and the pair of them, Einstein and Bohr, became the gurus of the nuclear fraternity. Bohr's position was helped by his government's support, proud to have a Nobel laureate in the country. He also benefitted from the financial support of the Carlsberg Foundation, brewers of the famous lager. He was able to establish a nuclear centre of excellence, now the Niels Bohr Institute, and invite prominent physicists of the day to come to the Institute for a long stay. Even after the Second World War, Copenhagen was the go-to place for sabbatical leave. During the period from about 1920 to 1939, the trains to Copenhagen were always full of scientists, everyone went to Copenhagen to ask Bohr's opinion on the many new ideas that were buzzing around.

Bohr also became a saviour for many of the Jews who were trying to escape from the Nazi regime. Many scientists with Jewish connections fled to Copenhagen since it became a well-known clearing house to find refugee jobs in the USA, with the help of funds from the Rockefeller Foundation. In the late 1930s London was also a clearing house, with a convention at the Albert Hall devoted to finding positions for anyone fleeing Germany, not just scientists. When Hitler occupied Denmark, the Gestapo were somewhat lenient to start with because they wanted to retain the Danish food production of pork and dairy products for the benefit of the Wehrmacht. Bohr had a Jewish mother and was always in fear of his position. Nevertheless, he supported all the refugees who came his way. Two Nobel prize winners asked him to take care of their Nobel prize gold medals, since the Nazis would confiscate anything made of gold. He agreed to do so, but when the Gestapo started to be less lenient, he decided to hide the medals by dissolving them in acid, and they sat in a bottle of dark brown liquid on a shelf in his laboratory until the end of the war. After the war the gold was recovered, and the medals re-issued.

The British wanted Bohr to escape to England and help with the war effort. They sent a secret message, via neutral Sweden, in the form of a microfiche dot inserted into the end of a large key, but Bohr did not want to desert his homeland and decided to remain in Denmark. He refused the invitation and replied by means of another microdot encased into a filling in a tooth of the courier. At this time, 1941, escapees from Germany were trying to get into Sweden by boat but were refused permission to land. Eventually, Bohr decided he needed to get out of Denmark for his own sake and went to Sweden where he used his influence to persuade the King to allow refugees to land before leaving himself. He flew to England in the bomb compartment of a Mosquito bomber and passed out because he didn't hear the instruction to switch on his oxygen supply. He then went over to the USA but did not play a significant part in the Manhattan Project.

Otto Frisch had escaped from Germany himself and was now working in England with Rudolf Peierls at the University of Birmingham. When the possibility of a nuclear bomb was suggested it was believed it would have to be too large to deliver, certainly by an aircraft, but theoretical work by Peierls and Frisch resulted in their famous Memorandum giving a sound theoretical basis for believing that just a few kilos of U^{235} would be sufficient to make a bomb. The Memorandum was sent from the UK to Washington as a top-secret document. The US civil servant who received it was so over-zealous in his responsibility to protect it that he put it in a safe, where it languished for several days before being seen by Fermi and his US colleagues. This signalled the start of the Manhattan Project.

5.3 The Fission Process

The fission process results in the emission of two, sometimes three, large fragments. It is a bit like an axe cutting through a log, it seldom cuts exactly into two halves. It also releases numerous high energy gamma rays, of the order of 10, and several neutrons, about 2.4 on average, with energies in the MeV range. The most probable neutron energy is about 0.7 MeV, and the average neutron energy is about 2 MeV. The neutron energy spectrum for thermal fission of U^{235} is shown in Figure 5.2.

The shape, known as the Watt spectrum, is similar for other fissile nuclei. The neutrons emitted in a reactor will survive and experience different types of reaction as they lose energy. Many will collide and lose energy, but still survive and continue to lose energy in collisions until they finally exist in thermal equilibrium with their surroundings, bouncing around and sharing energy with their neighbours. These are referred to as thermal neutrons and take up the distribution of velocities found in normal matter; a distribution known as the Maxwellian distribution. Typical Maxwell distributions are shown in Figure 5.3.

Most of the neutron lifetime is spent at thermal energies since the slowing down at high velocity takes very little time. To avoid the complexity of having to consider a whole range of energies over the thermal Maxwellian spectrum it is common to use the most probable neutron energy, which is 0.02530 eV at 293.59°K, which converts to a neutron velocity of 2,200 m/s. The energy range of interest in reactor calculations is

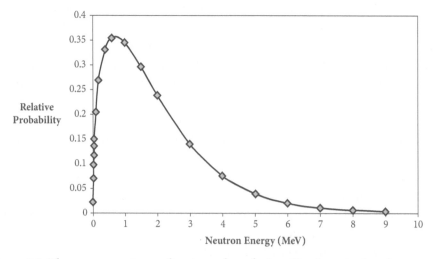

Figure 5.2 The energy spectrum of neutrons from fission. The shape is given by an empirical fit of the data to a mathematical expression known as the Watt spectrum

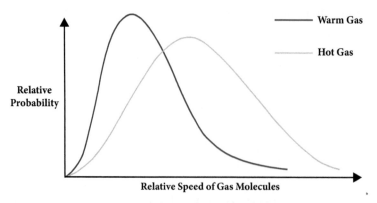

Figure 5.3 Typical Maxwell distribution of velocities

from 10 MeV down to 2.5×10^{-2} eV, an enormous range, giving the possibility of many different types of reaction in the neutron's slowing down history.

The three Fs in reactor physics are Fissile, Fissionable, and Fertile. **Fissile** are the nuclei that are capable of fission with neutrons at any energy, they are usually the odd mass nuclei in the actinides with atomic numbers above 89. There are exceptions and, when you are made aware that plutonium has as many as 20 identified isotopes, it is not surprising that some of the extremes break the rule. The well-known fissile materials are U^{235}, Pu^{239}, and U^{233}, and for thermal neutron energies the fission cross-sections can be several hundred barns. The even mass nuclei are **Fissionable**, meaning they are capable of fission but the cross-sections for thermal neutrons are extremely low. Energies in the MeV region are needed to cause significant fission. The term **Fertile** relates to nuclei that are capable of giving birth to fissile material by means of neutron capture—breeding. An example is U^{238}, which captures a neutron to create U^{239} (half-life 23 m) which then decays through Np^{239} as a short-lived (half-life 2.36 d) intermediary, before ending up as the long-lived isotope Pu^{239}, a fissile material.

The fission cross-section for U^{235} over the whole energy range is shown in Figure 5.4.

At high energy, it is about 2 barns. At 10 keV, as we enter the resonance region where the cross-section fluctuates violently, it ranges between 4 barns and 800 barns. Then, as we drop into the thermal region it rises to 1,000 barns at 0.01 eV. The fission cross-section for U^{235}, averaged over the thermal spectrum is 502 barns. Since the cross-sections vary so much with energy an accurate calculation would need to have a value for the cross-section at each energy over the whole range. This is not possible in practice and the solution is to break down the energy variation into regions and use the average for each region. A simple split would be to have three groups, a high-energy group based on the fission neutron spectrum, an intermediate group which would cover the resonance region in the case of U^{235}, and a thermal group. More than three groups are often defined when higher accuracy is needed. At low energies many cross-sections are proportional to the inverse of the velocity, suggesting that the length of

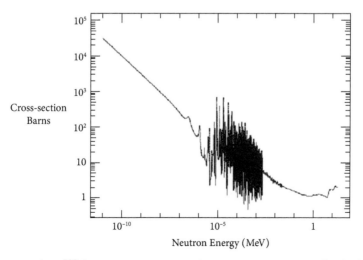

Figure 5.4 The U^{235} fission cross-section. The resonance region is clearly shown

time a neutron spends in the vicinity of a nucleus determines the cross-section. Think about it like a steel ball passing a magnet. At high speed it is unlikely to be captured and the probability of capture increases as it slows down and spends more time under the influence of the magnetic force.

5.4 Neutron Interactions

High-energy neutrons can enter a nucleus, say chromium in stainless steel, and exit leaving the nucleus in an excited state of an MeV or so. When this happens, the neutron leaves a large amount of energy behind in a single interaction. The excited chromium nucleus quickly decays to its ground state by emission of a gamma ray. This process is called **inelastic scattering,** which helps the neutron to lose a large amount of energy in one lump and proceed to thermal energy more quickly.

If the neutron simply bounces off the chromium nucleus, like a billiard ball, the chromium will recoil, and the neutron will shoot off at an angle. The energy loss by the neutron in this case depends on the angle of scattering and is at a maximum in a head on collision. The neutron loses very little energy by bouncing off a heavy nucleus but can lose a significant amount when it collides with a light nucleus such as hydrogen, deuterium, and even carbon and oxygen. This process is called **elastic scattering.** Nuclei that produce a significant loss of neutron energy by elastic scattering are called **moderators.** The percentage energy loss for a neutron in a head on collision with a few nuclei is given in Figure 5.5.

Neutrons can die and disappear through (n.p), (n,α), and (n,γ) reactions; called **capture reactions** with (n,γ) being by far the most probable. Neutrons often 'die' because they are captured by another nucleus and most of these cases are immediately followed by the emission of a gamma ray. Figure 5.6 gives the thermal neutron

H^1	D^2	C^{12}	O^{16}	Fe^{56}	Pb^{208}
100%	89%	28%	22%	7%	2%

Figure 5.5 Percentage energy loss by a neutron in a head on collision with various nuclei

B	Na	Mg	Al	Cr	Mn	Fe	Co	Ni	Cu
3837	0.417	0.06	0.189	2.72	11,79	2.29	32.98	3.93	3.36

Zr	Cd	Xe^{135}	In	Sm^{149}	Gd^{157}	Lu^{176}	Hf	Au	Ph
0,18	2918	2.6×10^6	202	74500	2.57×10^5	3048	92.3	88	0.158

Figure 5.6 Approximate thermal neutron absorption cross-sections for various nuclei. Cross-section in barns

capture cross-sections for some well-known elements. The ones with extremely large cross-sections are often referred to as **poisons**.

5.5 The Fate of Gamma Rays

Gamma rays play a big part in the energy produced by fission. They occur at the time of fission and later as a secondary release from neutron capture and radioactive fission products. Gammas lose their energy to heat through three main processes. **pair production**, which can only happen if the gamma ray has enough energy to produce an electron/positron pair; 1.02 MeV. This process is quite likely for high-energy gammas. The second method is through **the photoelectric effect**, where the gamma ray kicks an electron out of its atomic shell, often the inner shells, the K or L shells. When this happens, the electron produced loses energy to heat by scattering but the vacancy in the K shell, say, is filled by an electron from a higher shell and this generates an X ray. So, although the initial gamma has disappeared it is replaced by a photon of lower energy which, in turn, will undergo other gamma interactions before ending up as heat. Some gammas will escape from the reactor resulting in an undesirable loss of energy from the commercial point of view, but most will undergo energy loss within the core of the reactor.

The third process is called **Compton scattering**, discovered by Arthur Compton in 1923. It is illustrated in Figure 5.7. The photon can be considered as a particle, bouncing off the electron, and hence it loses energy to the recoiling electron which will eventually dissipate its energy in the form of heat. This process does not remove the photon, it simply reduces its energy and allows it to continue its journey that will eventually convert all its energy into heat.

There are several nuclear reactions which absorb photons, and some of them create neutrons and can be used as a useful neutron source:

Deuterium, H^2, produces a neutron when the photon energy is above 2.23 MeV:

$D + \gamma = n + p$, $Q = -2.23$ MeV; the deuteron splits into a proton and a neutron.

The other notable neutron source is Be, which requires a gamma ray above the 1.67 MeV threshold.

$Be^9 + \gamma = n + He^4 + He^4$ with $Q = -1.67$ MeV (Be^8 splits up into two alphas)

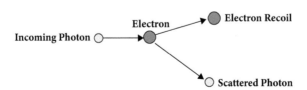

Figure 5.7 Compton scattering, a billiard-ball type of collision between a photon and an electron. The scattered photon has a longer wavelength and smaller frequency. It has less energy, lost to the electron recoil

5.6 Fission Fragments

The fission fragments for U^{235} have a mass distribution shown in Figure 5.8. The two peaks are in the region of Mass = 95 and Mass = 137, with Sr^{94} and Xe^{140} being an example of a pair of fragments. There is a very deep valley between the peaks showing that the fragments seldom have equal mass. The peak-to-valley ratio for fission in U^{235} is about 600 to 1. This form of two peaks is similar for other fissile nuclei but the valley fills in as the energy of the neutron causing fission rises, and at very high neutron energy the valley disappears and there is just a single peak. Most of the fission fragments are radioactive and will have their own decay chain through other radioactive nuclei. They will release a significant amount of energy, mainly through beta and gamma decay, and this will build up in a reactor to the extent that if a reactor is suddenly shut down this residual decay heat is about 6.5% of the power at shutdown.

5.7 Delayed Neutrons

Most of the fission products are radioactive and some are neutron precursors, meaning they decay to nuclei that decay by neutron emission. There are about 261 precursors producing neutrons with an average energy of 0.4 MeV. This produces delayed neutrons as distinct from the prompt neutrons emerging when fission takes place. There are six main groups of delayed neutrons, as detailed in the table shown in Figure 5.9. The six groups were originally suggested to enable the detailed time history of neutrons in the reactor to be calculated: reactor kinetics. They are grouped in time sets but the reality is that several precursors contribute to each group. Group 2 has contributions from I^{137} and Br^{88}. The energy of delayed neutrons varies between 0.3 and 0.9

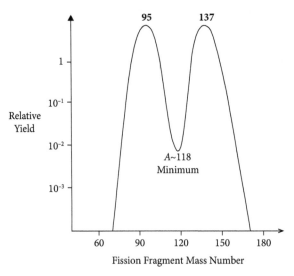

Figure 5.8 The fission product mass distribution from thermal neutron fission in U^{235}

	Mean Energy Mev	Average half-life of the group [seconds]			Delayed neutron fraction [%]		
		U^{235}	Pu^{239}	U^{233}	U^{235}	Pu^{239}	U^{233}
1	0.25	55.72	54.28	55.0	0.021	0.0072	0.0226
2	0.56	22.72	23.4	20.57	0.140	0.0626	0.0786
3	0.43	6.22	5.60	5.00	0.126	0.0444	0.0658
4	0.62	2.3	2.13	2.13	0.252	0.0685	0.0730
5	0.42	0.61	0.618	0.615	0.074	0.018	0.0135
6	–	0.23	0.257	0.277	0.027	0.0093	0.0087
				Total	0.64	0.21	0.26

Figure 5.9 Details of the six delayed neutron groups. More than one radioactive precursor may contribute to each group

MeV across the groups, which means they have a different effect on the chain reaction when compared to the more energetic prompt neutrons. They don't travel as far and are less likely to leak from the reactor.

The parameter to define delayed neutrons is the **delayed neutron fraction**, β, which, as the name suggests, is the fraction:

$$(\text{Delayed Neutrons})/(\text{Delayed} + \text{Prompt Neutrons})$$

However, the parameter of interest is the **effective delayed neutron fraction**, β_{eff}, which is β modified by the **importance factor**, I,

$$\beta_{eff} = \beta.I$$

The importance factor reflects the difference in energy between prompt and delayed neutrons. It can be slightly greater than 1.0 or slightly less depending on the type of reactor. A typical figure for the delayed neutron fraction in a PWR is $\beta_{eff} = 0.006$, whereas a fast reactor figure is about $\beta_{eff} = 0.0035$.

5.8 The Energy of Fission

The total energy released in fission is just over 200 MeV. A little bit more for Pu^{239} than U^{235}. In the case of U^{235} about 165 MeV comes from the kinetic energy of fission fragments, which is quickly turned into heat. The prompt neutrons, about 2.4 per fission, with an average energy of 2.0 MeV, contribute 4.8 MeV. The prompt gammas, released when fission takes place, account for an additional 8 MeV. The delayed energy from fission products amounts to 27 MeV but about 10 MeV is lost altogether in the escape of anti-neutrinos. There is also an additional 12 MeV available as heat when the prompt neutrons are captured inside the reactor and emit a gamma ray.

5.9 Decay Heat

Decay heat deserves a special mention because it is the main character in the drama associated with major reactor incidents. I am talking about a loss of coolant or a loss of power to drive the primary coolant pumps. This scenario is usually considered to be the most serious cause, top of the list, leading to a major incident. It is not difficult to shut a reactor down by inserting control rods, but the residual heat in the fuel cannot be switched off. It is the accumulation of all the radioactive fission products produced during normal operations. They have various half-lives, and they will remain until they have decayed away. Decay heat is typically about 6.5% of the heat when the reactor is up and running. For a PWR 6.5% of 3,500 MW (th), the decay heat on shutdown is 227.5 MW. This very large amount of residual heat diminishes over minutes, hours, and days as shown in Figure 5.10. Dealing with this formidable problem is discussed in the chapter on safety (Chapter 8).

5.10 The Chain Reaction

When it was understood that the fission process produced several neutrons, the possibility of a chain reaction was immediately recognized. Not long after the discovery of fission it was realized that it was only the U^{235} isotope that played a part in fission and that the fission process was much larger with thermal neutrons. Was a chain reaction possible? The only way to find out would be to build an experimental pile containing uranium and a moderator and hope that a chain reaction could be sustained and controlled.

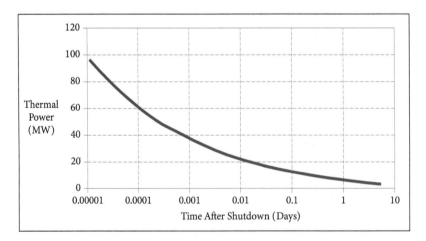

Figure 5.10 Nuclear decay heat reduction as a function of time after shut-down

The challenge to prove that a chain reaction was possible would have to be by using natural uranium, containing only 0.72% U^{235}, and a carefully designed experimental pile. The trick would be to get the neutrons down into the thermal region without losing too many of them to capture. Out of 2.4 released during fission it would need at least one to remain and create the next link in the chain. The choice of materials would be the key to success.

In the early 1940s there appeared to be two favourites for the choice of a moderator, pure carbon, and heavy water. Carbon and deuterium were both acceptable as a moderator with very low capture cross-sections. Hydrogen would obviously be the best moderator, but it had a high capture cross-section which ruled it out for a reactor fuelled with natural uranium. Experiments to confirm the suitability of carbon were very favourable and the results were classified. The classification of scientific results was something new to all the scientists working in the field and many were unhappy about restricting publication since their careers and reputation were very closely linked to the publicity of their work. The Germans also carried out experiments on carbon but came to the wrong conclusion. They believed that it was not suitable as a moderator. Their carbon samples seemed to indicate that carbon had a large capture cross-section. The reason for this difference in opinion is unclear but may be due to some unrecognized impurities in the test samples used by the Germans. Their false results led to them to dismiss carbon and link their nuclear programme to the production of heavy water in Norway. The Americans were very aware that the presence of boron in graphite could not be tolerated, so they took steps to remove it during the purification process. The first nuclear pile successfully demonstrated that a chain reaction was possible and opened the door to the use of fission as a nuclear weapon in the first place and later as a source of Atoms for Peace.

The fission process in other nuclei is very similar to U^{235}, with relatively minor changes in the numbers. Figure 5.11 gives some of the most significant parameters for fission in various nuclei.

Fission Isotope	Fission Cross-section 0.025 ev/2.0 MeV	Prompt Neutrons 0,025 ev/2.0 MeV	Delayed Neutrons 0,025 ev/2.0 MeV
U^{235}	585/1.27	2.42/2.63	0.0162/0.0165
U^{238}	0.000027/0.57	2.36/2.60	0.0478/0.0478
U^{233}	531/1.98	2.48/2,63	0.0067/0.0077
Pu^{239}	747/1.93	2.87/3.16	0.0065/0.0067
Pu^{241}	1012/1.76	2.92/3.21	0.0160/0.0160

Figure 5.11 Fission cross-sections for thermal and 2.0 MeV neutrons. Numbers of prompt and delayed neutrons from fission by thermal and 2.0 MeV neutrons

6

Nuclear Reactors in General

6.1 Nuclear Reactor Calculations

We have just dealt with the characteristics of fission and can now apply this knowledge to give us a good understanding of what is going on in a real reactor. Nuclear reactors and chain reactions were new in 1939, so let us take a careful look at what we are dealing with. A modern reactor, 3,500 MW (th), requires about 10^{20} fissions per sec to generate the energy. The body of the reactor is a sea of neutrons, a veritable ocean, which immediately brings to mind concepts related to a statistical approach. Terms like neutron flux spring to mind. Neutron flux is simply defined as the number of neutrons passing through a 1 cm² area per sec, regardless of direction, which makes it a scalar quantity. An analogous situation, when dealing with a statistically large number of 'particles', is the case where a cup of salt is emptied into a bath of water and, after it dissolves, it moves around by diffusion. The movement of neutrons or salt is dealt with by diffusion theory, a well-established theoretical approach based on Fick's law, which simply states that neutrons will diffuse from a high concentration to a low concentration, and a continuity equation, which states that the neutron density is equal to the birth rate of neutrons, minus the absorption rate, minus the leakage rate.

An alternative treatment is the well-known transport theory, based on Boltzmann's equation. It describes the statistical behaviour of a thermodynamic system not in equilibrium. This approach can be used to define the motion of heat as it travels down a copper rod towards the cold end, a transportation of energy. Transport theory and diffusion theory have both been used to calculate the properties of reactors with immediate success and provide a good insight into the behaviour of a critical assembly. They are excellent for simple geometries with specified boundary conditions but must adopt approximations when dealing with a complex array, such as a cube of graphite with cylindrical rods of uranium inserted, not to mention the complexity of cooling loops.

Fortunately, computer power has increased so much that complicated geometries can be defined in a computer model and the Monte Carlo method, described in Section 6.6, is by far and away the most reliable and accurate way of dealing with calculations to obtain everything we need to design a reactor.

6.2 The Growth of the Neutron Population

From the moment that fission was discovered in 1938, with the realization that several neutrons were emitted during the process, physicists started to think about the possibility of a chain reaction. If we consider a nuclear reactor operating in a stable

Understanding Nuclear Reactors. Brian Hooton, Oxford University Press. © Brian Hooton (2024).
DOI: 10.1093/oso/9780198902652.003.0006

condition and take a snapshot of the core, we are witnessing the neutron population at a point in time. It is a pure number, not a rate per second or a statement on the neutron flux passing through an area. The growth of this number is what we are going to examine. In formal terms, the way the neutron population changes with time is referred to as reactor kinetics.

The neutron population in a reactor has an analogy with the human population in a country. The growth is determined by the balance, or imbalance, of the birth rate, and the death rate. The difference between humans and neutrons is the time between generations. In the human case it is many years, but in the case of a reactor it is a fraction of a second. The population in a country grows slowly but the neutron population in a reactor can grow at an alarming rate.

A constant neutron population seems to reflect the fact that the neutron birth rate equals the death rate, but this is only true if there are no additional neutron sources, such as Cf^{252}, which may be in the reactor to facilitate a smooth growth of the neutron population during start-up. In the human analogy, additional humans may be thought of as the regular arrival of 'boat people', which will slowly increase the population. We will now look at the two cases of growth, the simple case with no additional source and the other case where growth takes place with an additional source present.

To examine and understand what is happening in reactors when the birth rate exceeds the death rate requires the definition of just two parameters. The first is the concept of a time between generations, denoted by g, which is the average time between a neutron being born and giving birth to more neutrons. The time between generations for prompt neutrons, those that are born at the moment of fission, depends on the composition, geometry, and temperature of the reactor. It takes about 20 collisions to thermalize neutrons in hydrogen, whereas in graphite, it would take just over 100. Experiments, using short nanosecond pulses of neutrons, can be used to determine the actual thermalization times in real cases. The results show that the time between generations, ignoring delayed neutrons, can be of the order of milliseconds, which would result in a very rapid growth of the neutron population.

The second parameter is the neutron multiplication factor, k, defined as the number of neutrons in a generation divided by the number in the previous generation. This factor depends only on the composition, geometry, and temperature of the system. The original definition of k referred to an infinite system with another parameter k_{eff} being the value of k in a finite system, with neutrons escaping at the boundaries. This distinction is not really necessary since we always deal with finite systems so just one parameter, k_{eff}, will be used here for the change in neutron population per generation in a finite system.

Consider a simple uniform system, without any additional sources to complicate the issue. When k_{eff} is exactly equal to 1.0 then the population is stable and does not change with time. If k_{eff} is not equal to 1.0 then the initial population, $N(0)$, taken as a snapshot at time zero, will grow (or decline) to $N(0) k_{eff}$ after the first generation then $(N(0) k_{eff})$ k_{eff} after the next generation becoming $N(0) k_{eff}^n$ after the nth generation. This reflects an increasing population if k_{eff} is greater than 1.0 and a declining population if k_{eff} is less than 1.0. Using g as the time between generations means that after a time t we will have had n = t/g generations and the population after time t is given by:

$$N(t) = N(0)k_{eff}^{t/g}$$

If k_{eff} is very close to 1.0, then we have an exponential growth or decline, growth when k_{eff} is greater than 1.0 and decline if k_{eff} is less than 1.0.

$$N(t) = N(0)\exp((k_{eff} - 1)t/g).$$

If k_{eff} is greater than 1.0 this allows us to state a doubling time of $0.693\ g/(k_{eff} -1)$ and, if we take values of g = 0.1 sec as a reasonable time between generations, including the effect of delayed neutrons, for a U^{235} system, and k_{eff} = 1.004 to reflect a reasonable growth rate the doubling time is 17.3 s.

The three situations that may exist in a reactor are when k_{eff} = 1.0 with the population constant, referred to as **critical**. When k_{eff} is less than 1.0 and the population is in decline, it is called **sub-critical**, and when k_{eff} is greater than 1.0, it is called **super-critical**.

Now consider the effect of an additional source of neutrons, S neutrons per s to a reactor with a stable population and k = 1.0. The stability is disturbed by the entry of additional neutrons. They represent an increase in the birth rate and the population will rise. Yes, even though k_{eff} = 1.0 the population rises when an additional source is introduced. Under these circumstances, the three situations develop according to the following growth rates.

When k_{eff} = 1.0, the population increases linearly at a rate S.

When k_{eff} is less than 1.0 the population increases and reaches an equilibrium value, a plateau, as t approaches infinity; the equilibrium value is given by:

$$N(\inf) = S\ g/(1 - k_{eff}),$$

for example, when g = 0.1 and k_{eff} = 0.996 then the plateau $N(\inf) = S \times 25$.

When k_{eff} is greater than 1.0 the population increases very rapidly.

If a reactor is starting from a very low neutron population with the presence of a start-up source, S neutrons per s, and the value of k_{eff} is just less than 1.0 then the population will slowly rise to a plateau of $N(\inf) = Sg/(1 - k_{eff})$. If k_{eff} is then increased further, but remains less than 1.0, the population will again rise to a higher plateau and this process will continue until k_{eff} is greater than 1.0 when the reactor becomes super-critical, and the population continues to rise exponentially. This is exactly how Fermi started his first reactor CP1 in Chicago (see Section 7.2).

In a real power reactor, the start-up source strength is insignificant compared to the operational neutron population and we consider the reactor to be critical and stable when k_{eff} = 1.0. It is quite evident that k_{eff} can never be precisely 1.0 because it doesn't matter how many decimal places you consider, k_{eff} is always above 1.0 or below 1.0.

A stable reactor operates with k_{eff} oscillating slightly above and slightly below, with an average value of $k_{eff} = 1.0$.

6.3 The Six Factor Formula

The value of k_{eff} can be estimated using the six factor formula, where each factor has a meaning; they are.
Thermal fission factor:

$$\eta = \text{(neutrons produced from fission)/(absorption in fuel isotope)}$$

Thermal utilization factor:

$$f = \text{(neutrons absorbed by the fuel isotope)/(neutrons absorbed anywhere)}$$

Resonance escape probability:

$$p = \text{(fission neutrons slowed to thermal energies without absorption)/}$$
$$\text{(total fission neutrons)}$$

Fast fission factor:

$$\varepsilon = \text{(total number of fission neutrons)/}$$
$$\text{(number of fission neutrons from just thermal fissions)}$$

Fast no-leakage probability:

$$Pf = \text{(number of fast neutrons that do not leak)/}$$
$$\text{(number of fast neutrons from all fissions)}$$

Thermal non-leakage probability:

$$Pt = \text{(number of thermal neutrons that do not leak)/}$$
$$\text{(number of thermal neutrons from all fissions)}$$

k_{eff} = The product of all six factors, and typical values for a thermal reactor are:

$$k_{eff} = 1.65 \times 0.71 \times 0.87 \times 1.02 \times 0.97 \times 0.99, \text{ which gives } k_{eff} = 0.9983$$

6.4 The Effect of Delayed Neutrons on Reactor Control

Let us suppose that a reactor is running at a value of $k_{eff} = 1.004$, then the doubling time takes about 180 generations. If we consider only the prompt neutrons from fission, with a generation time of a millisecond, we will experience a doubling time of only 0.18 seconds, which is difficult to manage. Fortunately, we know from our fission characteristics that there is a small percentage of delayed neutrons arising from the radioactivity of the fission products. Some of these neutrons are still around after tens of seconds, and they can be used to effectively extend the generation time. Consider the situation where the prompt neutrons alone would yield a k_{eff} value just less than 1.0, i.e., sub-critical. This would not, in itself, cause a rapid rise and we can benefit from the fact that, when we take the delayed neutrons into account, the delayed neutrons will take k_{eff} just above 1.0 with the distinct advantage of a much longer time between generations. In a general sense, the delayed neutrons can extend the generation time by about a factor of 100. This takes the doubling time, for a value of $k_{eff} = 1.004$ up to 18 sec, which is much more manageable. Nevertheless, there should be no hurry to increase the power of a reactor and operators will always keep the value of k_{eff} very close to 1.00. The effect of delayed neutrons on slowing down the growth of power in a reactor is so important that it is worthwhile expressing it again in a slightly different manner.

With a reactor operating in a sub-critical state on prompt neutrons alone, the chain reaction would tend to die out, but the delayed neutrons come along a moment later, just in time to sustain the chain reaction, and enable exponential growth at a rate that is comfortable for control.

6.5 Reactivity

A more common parameter for defining and discussing the growth of the neutron population is **reactivity**. It is probably the most important parameter in nuclear parlance since it is used to define the change in the state of a reactor, and crops up in all aspects of reactor control. It is given the symbol rho (ρ) defined as:

$$\rho = (k_{eff} - 1)/k_{eff}$$

when $k_{eff} = 1$, and the reactor is stable, the reactivity is zero.

Reactivity can be viewed as the departure of k_{eff} from unity, Δk_{eff}, to specify the decline or growth of the neutron population.

The values of reactivity need to be defined in units with names for a common understanding. Unfortunately, quite a few names and units exist, and this can lead to unnecessary confusion.

The simplest way of dealing with reactivity is to view it as a percentage of k_{eff}. We can define a 1% value and give it a name, but since this is a large value, we will also use the **per cent mille**, pcm, which is a thousandth of 1%. In the UK reactor parlance, the

1% value is called a Nile. This unusual name stems from the fact that it is considered to be a large value of Δk_{eff}, and the Nile is a large river with a large delta! The pcm value is called a milliNile in the UK. The Nile is not an internationally accepted unit. Other countries use other names with the French sticking to percentage, but the USA have complicated the issue by being more rigorous. They do use pcm for most situations but also use another unit, the Dollar, which is somewhat different and requires a formal definition.

The Dollar can be viewed as the difference between the conditions of prompt criticality and criticality using both prompt and delayed neutrons. This interpretation means that a change in reactivity of 1$ would result in a reactor going prompt critical. This varies according to the reactor conditions since the effective delayed neutron fraction β_{eff} can change from 0.007 at the beginning of a cycle to 0.005 at the end. The reactivity has been normalized to the delayed neutron fraction, which leads to the definition of the Dollar as:

$$\text{Dollars} = \rho/\beta_{eff}$$

For example, at the beginning of a cycle with $\beta_{eff} = 0.006$ and assuming a value of $k_{eff} = 0.99$ the reactivity in pcm would be −1,000, but the reactivity in Dollars would be −0.01/0.006, which is −1.67$ or if you prefer it, 167 cents.

This alternative unit, the Dollar, can only cause confusion, and so may the UK unit, the milliNile, so when I refer to reactivity, I will use the pcm unit for reactivity. A thousandth of 1% is self-explanatory.

Diffusion theory and transport theory have difficulty dealing with complex geometries but, fortunately, modern digital computers allow us to build a model to simulate a reactor, giving us an alternative way of understanding the complex activities taking place inside a real reactor. We can get reliable numbers without using mathematics. Modelling can result in a value for k_{eff} for a defined geometry and composition. This approach is also used to obtain a value for k_{eff} in experimental laboratory circumstances, to assure safety from a super-critical event.

In a summary of the theoretical options for dealing with reactors, the use of the three parameters, 'k_{eff}', 'reactivity', and 'time between generations' gives us an adequate way forward for discussing reactor operations. If this is combined with the computer modelling approach to predict the performance of a particular design, it is possible to get a good understanding of how reactors behave without the need to understand the mathematics of diffusion theory and transport theory.

6.6 Monte Carlo Models

We can get a good, reliable, and detailed picture of what is happening inside a reactor using modern digital computers. The first step is to set up a three-dimensional model of the reactor by specifying the geometry of all the regions in the reactor in fine detail. Tools are available to specify all kinds of shapes, and to give them three-dimensional coordinates. The chemical composition of each region can be specified, and clever procedures used to finally obtain a very accurate model of the system. Once the system has been defined a neutron source can be specified to set the ball rolling. The computer will track the neutron by predicting its first interaction on a statistical basis. This uses a data base containing all the cross-sections for neutron interactions. The computer will choose a direction, a distance and an interaction by 'throwing a dice' containing all the possibilities with the correct probability. In practice the computer chooses a random number (often several) and uses it to decide the first neutron interaction, and then, if the neutron survives, it follows it forward to its next interaction. A computer is so fast it can build up a complete picture of each encounter and retain a record of all the interactions that have taken place. These calculations are often referred to as Monte Carlo calculations, recognizing the similarity between choosing a random number in the computer and spinning the roulette wheel at Monte Carlo. The method simulates a reactor in operation and can determine the number of actinides and other fission products produced as the reactor continues to burn fuel.

The results of Monte Carlo models and even the results of mathematical models, where approximations have been made, can be validated by comparison with real laboratory critical assemblies. At the end of the day reactors are built with very adequate safety contingencies, with the operation of control rods, and other factors affecting reactivity, giving plenty of flexibility to operate the reactor and deal with all the uncertainties associated with the original design specification.

The plutonium produced in a reactor is of great interest to Euratom and the International Atomic Energy Agency (IAEA), who are the agencies involved in nuclear safeguards to monitor the non-proliferation of nuclear weapons. Monte Carlo calculations predict the amount of plutonium in spent fuel, and its isotopic composition giving the first 'book value' for the plutonium inventory at the time of discharge. If the fuel happens to be reprocessed then a sample of the dissolved nitrate can be taken and analysed to give the first accurate and reliable 'measured value' of the plutonium content of the discharged fuel.

6.7 Nuclear Reactor Operations

The **critical mass** is the smallest amount of fissile material needed for a sustained chain reaction. The value of the critical mass depends on all the materials in the system, the geometry, and the temperature. Whenever fissile material is being used, a criticality certificate is needed to authorize the safe operation of the activities. An area is defined and all the materials that may be used in the area must be specified to enable calculations of k_{eff} to be carried out under all reasonable circumstances. The geometry is very significant, and some facilities will design vessels and containers of fissile materials to be long and thin, a concept known as 'safe by shape'.

The nuclear material in a reactor will exceed the critical mass when the reactor is producing power, but the geometry and control rods will keep it safe all the time.

In the very early days of nuclear reactors natural uranium, with only 0.72% of fissile U^{235}, was the only source of fuel and careful neutron economy was needed to ensure that at least one neutron from fission managed to survive and create the next stage in the chain reaction. Neutrons would escape and be captured by other nuclei in the system and to maximize the probability of fission required slowing the speed of the neutrons down to thermal energies, taking advantage of the very high fission cross-section in U^{235} for thermal neutrons. The slowing-down process, known as moderation, required light nuclei to enable the neutrons to lose energy by collisions, with the qualification that the nuclei they encountered to do this must not capture them. These requirements resulted in the early reactors being moderated by heavy water or graphite (carbon). The reactors were classified as thermal reactors as distinct from the fast reactors that appeared much later. Thermal reactors continue to dominate the production of nuclear electricity. When enriched uranium became available reactor designs started to use normal water as a moderator and coolant but neutron economy remained an essential part of reactor design and thermal reactors remain the popular option because of their neutron economy characteristics.

The neutron flux in a reactor is not uniform, either radially or axially, and this will lead to different rates of burn up with commercial costs in fuel management. There are also operational problems if hot spots could damage the fuel pins. resulting in a release of fission products into the coolant. The flux profile depends on the design and the unavoidable presence of control rods, poisons, moderators, and the primary cooling system. Varying the enrichment across the core can be used to optimize the commercial aspects, as can moving the fuel around (fuel shuffling) as ongoing burnup starts to affect the profile.

Before discussing the time-dependent operations of a reactor in more detail, let us examine the five components that affect the operation and performance of a reactor.

6.8 Fuel

Uranium continues to be the main nuclear material used to create a chain reaction. It can be natural uranium with a U^{235} component of 0.72% or enriched uranium. Other artificially produced fissile materials such as Pu^{239}, Pu^{241}, and U^{233} also contribute to the fissile worth of fuel. The fresh fuel in the early reactors was uranium metal but the oxide, UO_2, has become the most common compound to be used as a fuel. Other compounds, such as uranium carbide and uranium nitride are attractive because they are so stable at high temperatures, and some liquid forms, capable of being added to the coolant, are now proposed. One of the early concepts, proposed in the 1950s, was the use of **TRISO** particles in high-temperature gas-cooled reactors. TRISO means TRistructural-ISOtropic fuel, an acronym that covers various designs. A typical fuel may consist of a core seed of UO_2 mixed with UC, surrounded by layers of carbon, pyrolytic carbon, silicon carbide, and a dense outer layer of pyrolytic carbon. Each layer covers a particular functional requirement. The inner layer will absorb the fission fragments and is protected by the pyrolytic carbon. The silicon carbide is a ceramic that will survive the highest temperatures in all reactors and will protect the fuel in the event of a core meltdown. The safety aspects of TRISO fuel suggest we are bound to use it but the commercial cost when the entire fuel cycle is taken into account means we may not. Other safety measures to deal with a meltdown have advanced so far that there is a case to say that TRISO, with its complex structure leading to extra costs, is not needed. Fuel manufactured in a form of 'pebbles' was introduced in the twentieth century as gas-cooled high-temperature designs were being investigated and new developments for future reactors can be expected. Oxide mixtures of uranium and plutonium, mixed oxide fuel (MOX), have been used to some extent and could also become more prevalent in the near future. The oxides are readily compressed into small cylindrical pellets. These are loaded into long narrow cylindrical pins made of stainless steel, zircaloy, or other metallic alloys, chosen for their nuclear and engineering properties. These pins can be mounted in clusters, separated sufficiently to allow a coolant to pass through, and the clusters mounted into **assemblies** for loading into the reactor.

6.9 Moderators

The best moderators are light nuclei, the lighter the better, which makes hydrogen, usually in the form of H_2O, the undisputed champion. Unfortunately, hydrogen has a relatively high neutron capture cross-section which prevents a chain reaction from developing in natural uranium. If natural uranium is the desired fuel, then it is necessary to avoid hydrogen and to look elsewhere. The two notable options in the early days of nuclear power were heavy water and carbon, in the form of graphite. Using graphite as a moderator means going up from mass 2 to mass 12; quite inferior, but manageable. With large quantities of enriched uranium not yet available, the use of graphite was seen as a minor disadvantage in the early reactors. It was used in the first military reactors in the USA and the UK to produce plutonium, as well as in most of the first-generation power reactors.

Deuterium in the form of heavy water, although superior as a moderator, had to be manufactured by electrolysis of water, which was expensive and time-consuming to produce in large quantities. Nevertheless, its superiority as a moderator was recognized and heavy water reactors, mainly of the CANDU type, have been very successful.

The technical solution to enable natural water to be used in a reactor is to use enriched uranium. The enrichment overcomes the significant neutron capture disadvantage of hydrogen. The enriched uranium required when water is used is at least 3%. Using natural water has the bonus that it serves a dual-purpose function as a coolant as well as a moderator. When we take a look at coolants to remove the heat from the reactor, water is very high on the list. We have confidence in water with centuries of experience in using water and steam. The overall picture, when we look at the compromises that have to be made in the choice of moderators and coolants is often expressed as, you can use whatever you like, as long as its water! Other light nuclei Be, Li, and F, that feature in molten salt coolants are also very effective as moderators.

6.10 Coolants

Coolants are required to remove heat from the core of a reactor and to transfer the energy into a form suitable for the generation of electricity. The most common vehicle to produce electricity is steam generation and a pressurized water reactor (PWR), operating at a temperature of 345°C is capable of converting the heat into electricity with an efficiency of 30%. It uses the thermodynamic process known as the Rankin cycle.

Although water is a very acceptable coolant it boils at 100°C under normal pressure. This is clearly much too low for the generation of electricity. The boiling point of water can be increased by increasing the pressure. Water boils when its vapour pressure equals the pressure of its environment. In practice, a typical PWR would operate at 155 Bar, which is about 155 atmospheres pressure, and the boiling point at this pressure would be 345°C. To improve the thermodynamic efficiency would require operating at a higher temperature and, obviously, a higher pressure. Unfortunately, when we get above 345°C any small increase in temperature requires a large increase in pressure and 155 bar is already high enough.

The quest to operate at higher temperatures and benefit from higher thermal efficiency requires something other than water. A gas is an obvious choice for operation at normal pressure. The two common options, CO_2 and helium, behave quite well at high temperature and both were used in several of the early reactor designs. In the UK, the advanced gas-cooled reactor (AGR) gave a satisfactory performance but the overall design was abandoned in favour of the PWR. One of the problems with any gas cooled reactor is getting sufficient mass of coolant through the core to give adequate heat transfer. It requires a very high velocity of gas transfer and in a complex engineering system with gas moving close to supersonic speeds vibrations can disturb the structure. We will see that helium coolant is proposed for some new reactor designs and one hopes that the technology required for a high mass transfer of gas has progressed, and improved, as we have now moved into the twenty-first century. High temperature He gas has been proposed as a coolant in several new reactor designs. This would open the door to using a gas turbine (the Brayton cycle) with an efficiency in the region of 50%.

Liquid metals have been used for several reactors in the twentieth century, notably lead—(MP 1,749°C, BP 3,270°C), sodium (MP 883°C, BP 980°C), and a eutectic mixture of sodium and potassium called NaK (MP −12°C, BP 785°C). A lead/bismuth alloy, and mercury, have also been investigated. These have all had teething problems and sodium experience at Dounreay with the Fast Breeder Reactor has not helped to raise confidence in the use of sodium. It is, after all, a very reactive material in the presence of air and moisture, which makes even a pin-prick leak in a weld a potential shutdown problem.

There is a very large list of options for molten salt coolants; many variations are under consideration for fourth generation reactors. In the twentieth century, the lithium, beryllium, and fluorine combination FLiBe (MP 459°C, BP 1,430°C) was used and encouraging results have resulted in optimistic views about the future of molten salt reactors (MSRs). The light elements present in molten salts also have a useful value as moderators.

6.11 Poisons

Any nucleus that has a high neutron absorption cross-section can be referred to as a poison. They kill neutrons and remove them from the chain reaction. Thermal fission cross-sections are often several hundred barns so anything greater than 1,000 barns tend to be the only ones referred to as a poison. Xe^{135} has a cross-section for neutron absorption at 2.7 million barns, which is exceptional. A selection of 'poisons' is given in Figure 6.1.

Poisons can be grouped into three categories.

A. Control poisons. Those that are used for reactivity control.
B. The unavoidable poisons that are the consequence of fission and appear in a reactor as the fission products progress down their radioactive decay chains.
C. Burnable poisons that are deliberately introduced into a reactor to prolong the length of time that fuel can remain in the reactor.

6.12 Control Poisons

There are many isotopes that have a high neutron capture cross-section and are candidates for killing chain reactions. Most of them are notable for their capture cross-sections in the thermal region but they can be combined in a cocktail to capture neutrons over a wide energy range. In the very first Fermi reactor cadmium sheets, nailed to wooden boards, were used, basic but effective control. B^{10} is a 20% constituent of natural boron, it has a high neutron capture cross-section through the B^{10}

Xe^{135}	Sm^{149}	He^3	B^{10}	Cd^{113}	Gd^{157}	Kr^{83}	Nd^{143}
2.65×10^6	40,140	5,333	3,837	20,615	254,000	197	325

Figure 6.1 A selection of poisons. Thermal (Maxwellian) capture cross-sections; barns

(n α) reaction leading to Li^7. Natural or enriched boron in the form of boron carbide, or as a constituent of stainless steel, is a very common control rod option.

PWRs use several control mechanisms. An alloy of silver, indium, and cadmium is a major cocktail to operate quickly in a shutdown operation because it is effective over a wide energy range. Boron is used in control rods but also in the form of boric acid, dissolved in the coolant. This gives another dimension to the options for controlling reactivity; it has several advantages. It is very intimately placed as the coolant flows up through the core to perform its primary function. It can be varied from large-enough concentrations at the start of operations to zero as the reactor approaches a fuel change. Changing the boron concentration is an ideal way of changing the reactivity by a small amount for operational purposes and it is readily done on a daily, or even hourly, basis.

6.13 Unavoidable Poisons

These are created soon after fission takes place and are present in the radioactive decay chains of fission products. They build up and die away as the decay chains evolve. Although there are many poisons in fission products the ones that stand out are Xe^{135} and Sm^{149}. These poisons, together with a multitude of other fission products, that are collectively known as 'lumped fission product poisons', can cause a large drop in reactivity.

Xenon Poisoning

Some Xe^{135} is produced directly from fission but most of it appears as a daughter product in the decay chain for mass 135 (see Figure 6.2).

The total fission yield of Xe^{135} from U^{235} is about 0.064. The half-life of Te^{135} is so short its contribution can be added to I^{135} and ignored as a separate source. As the amount of Xe^{135} builds up in the reactor it is constantly being removed by natural radioactive decay and by interactions with neutrons. The neutron absorption cross-section, 2.72×10^6 barns, is so high that, in a high flux reactor, the loss of Xe^{135} by neutron capture can be comparable to the loss by natural decay. This loss can be represented as an 'effective half-life due to neutron capture'. In a neutron flux of 10^{13} n cm^{-2} s^{-1} this half-life would be about 5.5 hr, a more rapid loss than the natural half-life of 9.2 hr. Calculations to give the equilibrium value of Xe^{135}, as a reactor builds up to a stable power, and calculations to determine the history of Xe^{135} after shutdown

$$Te^{135} \xrightarrow{\beta^- (11s)} I^{135} \xrightarrow{\beta^- (6.7\,hrs)} Xe^{135} \xrightarrow{\beta^- (9.2hrs)} Cs^{135} \xrightarrow{\beta^- (2\times10^6\,yr)} Ba^{135} \text{ (stable)}$$

Figure 6.2 The Xe^{135} decay chain

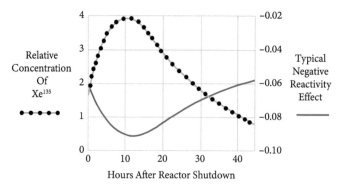

Figure 6.3 Time dependence for residual Xe^{135}. The initial rise is due to the formation of Xe^{135} from the decay of I^{135}. The right-hand axis shows the effect of Xe^{135} on reactivity

are readily achieved, with an example of the shutdown time dependence being given in Figure 6.3.

The Xe^{135} concentration is seen to rise immediately after shutdown since removal by neutron absorption is not taking place and Xe^{135} is still being created by the 6.7 hr decay of I^{135}. After shutdown, it can take several hours for the Xe^{135} amount to reach its maximum value and after about 20 hr the level will start to fall according to the natural 9.2 hr decay constant.

Samarium Poisoning

The effect of Sm^{149} as a poison is somewhat different from Xe^{135} because it is a stable isotope. The only way to remove it is by neutron capture to convert it into Sm^{150}. The direct production of Sm^{149} is negligible compared to that resulting from the decay of Ne^{149} and Pr^{149}. The decay chain is shown in Figure 6.4, and once again the contribution from the short-lived component, Ne^{149}, can be lumped together with the 54 h Pr^{149}. The U^{235} yield of Ne^{149} and Pr^{149} together is 0.0113. Sm^{149} has a smaller neutron absorption cross-section than Xe^{135}, so it has a much longer effective half-life due to neutron absorption. Since Sm^{149} is stable, the overall time effect for samarium as a poison is not the same as that for xenon poisoning. It takes about 36 days for the samarium to reach an equilibrium value in a flux of 10^{13} n cm^{-2} sec^{-1}, and when the reactor shuts down the samarium will build up, at the time constant for Pm^{149} until all the Pm^{149} has decayed. Samarium remains in the system after shutdown and, for a historical flux of 10^{14} n cm^{-2} sec^{-1}, the negative reactivity effect of samarium would be constant at about −1.5%. This is a very significant amount in terms of reactivity control.

The overall effect, of both xenon and samarium, is often referred to as **Xe poisoning.** Special procedures are necessary after shutdown because this poisoning becomes a maximum after about 10 hrs and leads to what is known as a **xenon-precluded start-up.** An immediate restart would require removing control rods to compensate

$$Ne^{149} \xrightarrow{} Pm^{149} \xrightarrow{} Sm^{149}$$
$$\beta^- \text{ (2 hrs)} \qquad \beta^- \text{ (54 hrs)} \qquad \text{(stable)}$$

Figure 6.4 The Sm^{149} decay chain

for the presence of Xe. This is totally unacceptable and dangerous, since it could contribute to an uncontrolled prompt-neutron chain reaction—an uncontained bomb!
The standard procedure is to wait until the Xe^{135} and other poisons have decayed to
an insignificant level. The long-term build-up of uncontrolled poisons eventually leads
to the need to replace the fuel, even though so little of the fissile material may have been
used.

In heavy water reactors, neutron capture in deuterium produces tritium, with a half-
life of 12 y, leading to He^3. This is also a poison with a thermal neutron absorption
cross-section of 5,330 b and a consequential effect on reactivity. Its presence cannot
be ignored. In practice, both tritium and He^3 have a commercial value and can be
removed from the coolant to make money and conveniently deal with the negative
reactivity effect of these isotopes at the same time.

6.14 Burnable Poisons

The purpose of a burnable poison is to create an initial negative reactivity in the fuel.
The fuel has a higher-than-normal enrichment to compensate for the presence of
the burnable poison. During the normal operation of the reactor, the poisons should
decrease their power at the same rate that the fuel's excess positive reactivity is diminished. This strategy should allow the fuel to remain in the reactor for much longer than
a fuel without a burnable poison.

The most common poisons in current use are boron and gadolinium. These can be
used in either a fixed form or as a fuel additive. Fuel additives complicate the fuel manufacture process and separate burnable poison rods are usually the preferred option.
Fixed poison rods can also be useful to tailor the flux and power across the geometry
of the reactor.

Non-burnable poisons, or perhaps one should say, almost non-burnable poisons,
also have a role to play in reactor design. Hafnium has five stable isotopes, Hf^{176} to
Hf^{180}, and all of them have a high neutron absorption cross-section, so one poison
leads to another, maintaining its effectiveness until Hf^{181} decays to the stable Ta^{181}.
There are many other options to consider, including erbium, lutecium, zirconium, and
even protactinium. This list demonstrates how sophisticated reactor core design has
become in the pursuit of improved performance and safety.

6.15 Engineering Materials

Metals and all other materials inside a reactor are chosen for their non-nuclear properties, their ability to withstand high temperatures, their corrosion resistance, and their

structural strength. They do have nuclear properties, which may well have to be considered when making a final choice of material to achieve a certain objective. The choice of fuel pin cladding is a good example of how materials have changed. In the early Magnox reactors, the cladding, essential to retain the fission products and protect the fuel from deterioration, was a magnesium alloy, **magnox**, with significant moderation properties since the mass of magnesium is only 24. Magnesium also has a very low thermal neutron absorption cross-section of only 0.059 barns. Magnox had properties that were ideal for a reactor that used natural uranium as fuel, but the corrosion properties underwater, when the spent fuel was kept in a storage pond, created problems. An important consideration for any material in a fuel assembly is how it is going to survive after discharge, when it may be sitting underwater in a spent fuel storage pond for more than 20 years!

The next obvious choice was stainless steel, and sure enough many reactors moved in this direction. Eventually an alloy of zirconium, zircaloy, was chosen for its overall properties, both nuclear and non-nuclear. This alloy is considered to be a better choice than stainless steel since it has a thermal conductivity 30% higher than stainless steel, facilitating the movement of heat from the fuel into the coolant. The linear coefficient of thermal expansion is about one-third of that for stainless steel, giving it superior dimensional stability at elevated temperatures. Zirconium ores would normally contain several per cent of hafnium which is a well-known neutron poison, so a nuclear grade has been developed with the hafnium content reduced to less than 0.01%. Small amounts of tin, niobium, iron, and chromium were introduced to improve the overall performance characteristics of zircaloy cladding.

Zirconium interacts with water at elevated temperatures to release hydrogen. This can diffuse into the zirconium itself resulting in hydrogen embrittlement, a well-known metallurgical problem, and it can also release free hydrogen with the danger of a hydrogen explosion. Fast neutrons can create voids in zirconium and other metals by a process of radiation damage. The fast neutrons may displace an atom as it recoils from a collision, and the recoil may end up in an interstitial position. Eventually, if this happens to a cluster of atoms close together, a void can form affecting the strength of the material. This is more of a problem with the high-energy neutrons in fast reactors.

The essential components of a reactor, fuel, moderators, coolants, engineering, and control materials are changing all the time with ever more stringent demands to improve safety and commercial profitability. Some of these changes are covered in Chapter 11 on the future of fission.

6.16 The Fast Reactor

The fast reactor is a very different animal to the PWR. Its characteristics were explored in the early days, but it only came into prominence with the concept of a reactor breeding more fissile material than it consumed; a breeding ratio greater than 1.0. The reality is that all reactors are breeders since they do produce some additional fissile material. Even the PWR produces large quantities of Pu^{239}, but since it is burned in-situ, and not harvested for later use, it is not seen as breeding. To achieve a breeding ratio of 1.0 requires a rethink about the fate of neutrons in a reactor. The concept of neutron economy dominated early reactor design, with thermalization needed to get the best out of U^{235} in natural uranium, but substantial breeding requires more neutrons to be absorbed by U^{238} and still maintain a chain reaction. An enriched fuel would be essential, and to prevent too much fission in U^{235} taking neutrons away from U^{238} capture, the design required a move away from the thermal region. The answer is to avoid a moderator and rely on fission at fast neutron energies in the region of an MeV or so. Fission does take place at this energy and even-mass nuclei like U^{238} and Pu^{240}, as well as all the other actinides, will make significant contributions. Moving away from thermal neutrons makes a big difference to all the cross-sections and we find ourselves in a completely different ball game. The change in cross-sections, when we move from the thermal to fast region, can be illustrated by the example that the ratio of the fission cross-section in Pu^{239} to the capture cross-section in U^{238} changes from about 100 down to 8 as we move from thermal to fast.

The fast breeder designs required a smaller compact core with higher temperature and energy density, no moderator, and an outside blanket for producing fresh fissile material that could be harvested by reprocessing and put back into the fuel cycle.

The high energy density and higher temperature requirements pointed towards liquid metal coolants such as sodium or lead, with a few other possibilities. The high energy density made it difficult to measure the power using conventional neutron detectors in the core because of the high temperature and gamma flux. An alternative method called the Campbell channel could be used. It was based on the noise in an ionization chamber, not the DC signal. Cambell developed noise theory in the early twentieth century to explain noise in the original radio valves. The variation of the root mean square (RMS) value of the noise turned out to give a reliable number for the neutron flux over a very wide range, even in the presence of gamma rays. Control of the fast reactor was also different without the resonance region to provide a negative fuel coefficient, although control was never a real problem.

The interest in fast breeders drifted away as the price of uranium fell, and producing electricity at the cheapest possible cost dominated the development of the nuclear industry. However, there are significant benefits in using fast neutrons, regardless of breeding, and these are emerging while Generation IV reactors are developing. The absence of a moderator is a great benefit and fast fission generates more neutrons. Using a metal coolant has the benefit of not requiring a pressurizer and gives a higher thermal efficiency with a higher temperature of operation. Sodium boils at 883° C, lead at 1,749° C, with molten salts now suggested as a possible alternative. The fast neutron cross-sections lead to a different population of the higher actinides in the core, with some benefits since most actinides provide additional fissile worth for fast neutrons, and calculations show that fast reactors produce a lower inventory of long-lived waste.

6.17 Hybrid Reactors

There have been many suggestions for the design of hybrid reactors. They are essentially reactors designed to have a sub-critical value of k_{eff} but operate in the presence of a strong external source of neutrons. An external source of interest is a large neutron source produced by a high-energy proton accelerator directed into a spallation source. Experiments to investigate this form of hybrid are taking place in Belgium, the MYRRHA LFR Project, as part of the EU Generation IV programme.

A second external source of interest is the 14 MeV neutrons generated by fusion. In this case, a sustained fusion reactor would have a sub-critical 'fission reactor' as part of the conventional blanket surrounding the fusion reactor and used to breed more tritium. This may sound to be a very bold suggestion but there are still uncertainties, some would say doubts, about the ability of a fusion reactor to be commercially viable operating alone!

These bold suggestions are examples of the wide range of proposals for making nuclear electricity in the coming decades. Chapter 11 goes into further detail about these possibilities.

7

Reactor Operations and Control

7.1 Controlling Reactors to Keep Them Safe

The public at large have an understandable fear about anything nuclear. It is difficult not to think about Hiroshima, and the horrific images of human suffering. We also have the memories of several 'nuclear disasters', particularly Chernobyl, so let us not be dismissive about their concerns, and take nuclear safety as the most important aspect of reactor management. The key word is control, which can be identified with the well-known parameter **reactivity,** so let us recap on how it relates to reactor control.

When the power in a reactor is stable the neutron population, which determines the power of the reactor, is also stable and the ratio of neutrons in one generation to the previous generation is equal to 1.0. The reactivity indicates the departure from this value so that a negative value means the neutron population, and power, is going down. If the reactivity is positive, it is increasing. There are many factors that affect reactivity when the reactor is up and running, control rods, poisons, fuel burnup, temperature, and changes in the density of anything in the reactor. You could say that any change in the reactor will affect reactivity.

7.2 The First Reactors

The first reactor was not controlled, it happened millions of years ago in central Africa. In 1972 scientists in a French laboratory were astonished to find that their samples from the uranium ore deposits in Oklo, Gabon, contained only 0.6% U^{235}, well below the norm. Investigations confirmed that the deposits were concentrated enough to support chain reactions and the production of fission products in an environment brought about by Mother Nature. This is the only natural reactor discovered so far and the next pile of uranium would not become critical until just after the discovery of fission in 1938.

The discovery of fission immediately raised the possibility of a chain reaction, but it would take time for the gaps in our knowledge to be filled. Early in 1939, at Columbia USA, measurements revealed that more than 1 fast neutron was emitted during fission. In the following year, U^{235} was identified as the only significant uranium isotope contributing to fission but hopes of a chain reaction in natural uranium remained optimistic as the estimates for the number of neutrons emitted had risen to 1.7; the actual value eventually turned out to be 2.4.

With such a low percentage of U^{235} in natural uranium, it was essential to have the best possible neutron economy in the reactor and graphite was the only moderator

Understanding Nuclear Reactors. Brian Hooton, Oxford University Press. © Brian Hooton (2024).
DOI: 10.1093/oso/9780198902652.003.0007

available to do the job. Heavy water was an excellent option but the cost, and time necessary to produce it in large quantities forced the scientists into the choice of graphite. The USA had some specialist knowledge of the production of graphite through the operation of the National Graphite Company, and boron was a well-known impurity. It would take two years of R&D before reactor-quality graphite, with almost zero boron, was available. The physics team in the USA, headed by Enrico Fermi, now had all the data to design and build a reactor with a good chance of producing a sustained chain reaction. They also now knew about delayed neutrons, which gave the operators much more time to stabilize the reactor as the chain reaction progressed.

The reactor built by Fermi in Chicago was his third attempt to produce a pile that would have a K_{eff} factor of 1.0. His first attempt, in September 1941 at Columbia, used uranium oxide contained in cubic metal boxes, but this gave a disappointing $K_{eff} = 0.87$. His next attempt used uranium stuffed into holes in the graphite, thus avoiding the metal plate boxes. This gave $K_{eff} = 0.918$—still not high enough.

7.3 Reactor CP1

The records show that the scientists involved in CP1 took a very responsible attitude to the risks involved in building the first pile of uranium and pulling out the control rods to make it go critical. Nevertheless, it must have been a worrying time with no one really knowing how the tests would turn out, particularly in case something surprising happened.

The pile consisted of 350 tonnes of graphite with holes drilled for the fuel and control rods. 45 tonnes of uranium oxide fuel were used in the form of pressed pellets, with some 5 tonnes of uranium metal added as the preferable metal became available.

The control rods were cadmium metal plates nailed onto wooden boards. The shutdown rod and emergency scram control, called the ZIP, could be manhandled, or cut by an axe to fall under gravity. It was not necessarily 100% safe, since in one modern power reactor a control rod got jammed as it was falling; you could say gravity failed! An additional safety feature was a bucket of cadmium nitrate to pour into the reactor as a last resort. One control rod was part of an automatic safety control system which would operate if the neutron count rate became too high. This did function during the morning of 2 December 1942, but not because the reactor was out of control: it was due to a setting on the neutron counter being too low. It interrupted the work and gave Fermi and his crew time out for lunch.

The neutron counters were specially made ionization chambers, Geiger counters if you prefer, containing some boron trifluoride gas. When a neutron enters the chamber it reacts with the boron via the B^{10} (n, α) Li^7 reaction. The α and the recoiling lithium create ionization and the electrons are collected at the anode of the counter. The final stage is to convert the electrical signal into the well-known audible click associated with a Geiger counter. The neutron counter(s) were within the core.

The start of a chain reaction in CP1 was initiated by neutrons from a radium/beryllium neutron source. Fermi could have left it to random neutrons from cosmic rays or spontaneous fission of U^{238} but it would have been haphazard, and having a definite start-up neutron source gives a smooth ride in the build-up of the chain reaction. All reactors use a start-up neutron source and there are many options, utilizing spontaneous fission by means of Cf^{252} or by (α, n) reactions. These are spaced throughout the core and may be removed after the reactor has been operational if their performance has been affected by neutron irradiation in the reactor. An antimony/beryllium source utilizing a (γ, n) reaction is also an option for a secondary source, with the Sb only becoming active after being irradiated in the reactor. It can have a useful life of many years operating as a start-up source.

Starting CP1 for the first time must have been a very dramatic moment. The procedure to withdraw the control rods and wind up the reactor was in the hands of Enrico Fermi. He knew exactly what to expect and withdrew the control rods slowly, a step at a time, with a dramatic pause at each stage to observe the effect. As a control rod is removed, the death rate for neutrons falls and the population of neutrons in the reactor increases. The counters click away and increase until they reach a plateau at the new stable rate. Pull out another control rod and the process repeats itself as the neutron flux moves up to the next plateau. After a while, the clicks become a continuous roar and it becomes time to switch over to a chart recorder. Even the range of the chart recorder needs to be changed as the test progresses, one step at a time. Eventually, the rise does not stop at a plateau, it continues a steep rise in the presence of a positive reactivity, a sustained chain reaction, and a definite need to take control. Fermi observed the climb for about 20 secs before stepping in and calling for the shutdown ZIP rod to be dropped in. The first power reactor; 0.5 watt.

7.4 Controlling Commercial Reactors

All power reactors operate in a similar manner to CP1 in start-up. They proceed, a step at a time, to a higher plateau as the power is increased. Day-to-day changes in reactivity due to operational demands require fine-tuning and we can now take a look at the control and operation of a reactor during its commercial life. Most of this chapter is presented in the context of the pressurized water reactor (PWR).

A nuclear reactor generates energy, which is processed through the generation of steam to the eventual product—electricity. To achieve this, the reactor is linked to a large number of very complex engineering systems consisting of pumps, heat exchangers, steam generators, pressure control units, condensers, steam dumps, and turbines. All of them are complex modules in their own right, requiring sophisticated control systems, and each is worthy of detailed description by a specialist, but we need this brief description of each component to understand how relevant they are to reactor control. The major components of a PWR are shown in Figure 7.1.

The pump drives the coolant into the reactor through the '**Cold Leg**', at a temperature in the region of 290°C. After passing through the core the coolant leaves at a higher temperature, 320°C, through the '**Hot Leg**'. It then passes into the steam generator

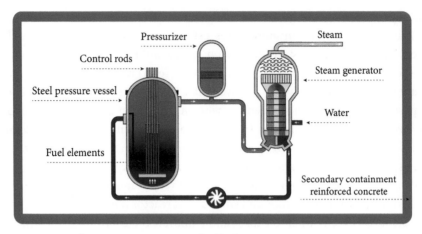

Figure 7.1 The main components of a PWR

where it exchanges energy with the secondary steam loop before leaving the steam generator via the '**Cross Over Leg**' and returning to the input for the primary pump.

7.5 The Reactor Pressure Vessel

The reactor pressure vessel (RPV), is at the heart of a nuclear power station. It contains the reactor core and has many input ports and exit ports to deal with the flow of coolant, the movement of control rods, and various instruments to measure the neutron flux. The coolant flow is easy to understand. It flows into the reactor through the cold leg and gets diverted down to the bottom of the vessel, where it then moves upward through the core, getting hotter all the time, until it leaves the pressure vessel through the hot leg. The pressure vessel is very large, often over 14 m tall and weighing over 400 tonnes which makes it a major engineering challenge when it comes to manufacture and installation. It is subject to the most rigorous quality assurance demands, but so is everything in the nuclear industry. A special-purpose mild steel is used to give strength to the main body of the pressure vessel, with an inner layer of stainless steel used where it comes into contact with the coolant. The vessel head is removable for refuelling and in a PWR, the control rod drive mechanisms are just above it (see Figure 7.1). The structure operates at high pressure in the case of a PWR and has the responsibility of protecting the environment from the highly radioactive nuclear core, it is the first line of defence. It is not surprising that during its construction and throughout its operational life the RPV is always under the microscope. The word integrity comes to mind when we question its adequacy for continuous operation over some 40 y. So, what could go wrong?

There are two main concerns, the first being a deterioration of its strength due to embrittlement caused by radiation damage, particularly neutron damage, since it surrounds the core. The second concern is the integrity of the many welds that support the entrance and exit ports. These can deteriorate over time and cracks may well develop. There are many possible causes for weld deterioration one of them being the presence of boron, often present in the coolant, which can lead to mild boric acid that will have some effect on any metal it encounters.

Fortunately, these problems are well understood, and inspectors do indeed keep the pressure vessel under a microscope for its entire lifetime. Tests can identify a weakness and a remedy actioned.

7.6 The Reactor Coolant Pump

These specialized spinning impeller pumps are an engineering marvel, a product of the nuclear age. They may weigh as much as 50 tonnes, and some require more than 5 MW of electricity to drive them. They are capable of pumping more than 5 tonnes of water per second. There will be several pumps, each driving an independent loop. If one fails, or is under maintenance, the remainder will be able to operate the reactor safely without requiring a shutdown. In a PWR the drive shaft of these pumps has to be taken through a special seal into the high-pressure region of the primary coolant. One side of the seal is at atmospheric pressure and the other side is at 155 Bar, the high pressure needed for water to exist as a liquid at over 300°C. To prevent leakage through the seal a separate stream of clean water, at a pressure just above 155 Bar, is injected into the seal. A very small amount will get into the coolant and the rest will be recycled for later use in the seal injection system.

7.7 The Pressurizer

PWRs require a special system to control the high pressure of the coolant. It can be thought of as a rather large kitchen kettle. It is partly filled with water and heated until it boils within its own pressure vessel. The pressure and temperature of the water can be adjusted to the required operational values by varying the heat, or by spraying cooler water into the system at the top of the pressurizer. Only one pressurizer is required to control all the cooling loops since they are all connected in a pressure sense. It has a pressure relief valve to deal with over pressurizing. See Figure 7.2.

7.8 The Steam Generator

There is a separate steam generator for each coolant loop. They weigh about 300 tonnes and convert the energy contained in the primary circuit to steam. The primary coolant enters a set of thin-walled tubes to facilitate the transfer of heat to the secondary circuit water. Steam is created and rises to the top of the generator where it passes through steam driers before proceeding on to the turbines. See Figure 7.3.

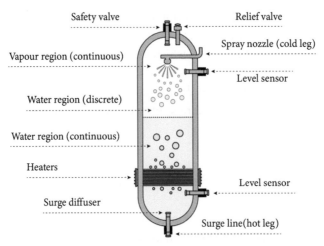

Figure 7.2 A PWR pressurizer. The bottom heaters and the top spray nozzle are used to control the pressure

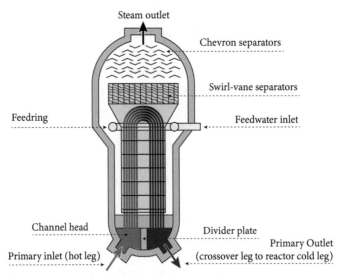

Figure 7.3 A steam generator. Chevron and swirl-vane separators are used for steam drying

7.9 The Boron Loading Loop

A PWR often uses a solution of boron, in the form of boric acid, as an additive to the primary coolant. It gives fine control of reactivity as the reactor progresses through its operational life and the core reactivity changes. A mini chemical plant is used to implement the changes in boron concentration. It is a loop used to clean up the coolant removed from the system by means of filters, and it deals with any radioactive contamination. The refreshed boron is blended with more boron or diluted with water to provide the new required concentration for feedback into the reactor. It should be noted that neutron reactions on B^{10} lead to tritium production which, over a passage of time, leads to He^3, a powerful poison with an absorption cross-section of 5,333 b. When boric acid is used to control reactivity in the coolant of a PWR, the build-up of tritium and He^3 needs to be dealt with.

The operation of this complex facility to raise steam and generate electricity has feedback effects with repercussions on the control of the reactor itself. Almost any change will have some effect on the reactivity in the core. This chapter looks at the many situations that the reactor operators have to contend with in their everyday operations to generate electricity. Although the various aspects are presented in the context of the well-known PWR, most of the principles are relevant to other types of reactors.

7.10 Power Measurement

The power produced in a reactor can be monitored by several means, giving the operator a complete picture of what is happening. Various types of neutron flux detectors will be positioned around the reactor to determine the reactor power. Unfortunately, they do not give a complete or accurate picture of the power, so let us have a minor digression and look at power measurement technology.

Neutron counters are an excellent and prompt indication of the neutron flux in the reactor, so you would expect them to give a reliable and accurate measure of the power, but inside the reactor the flux profile is complex and changes as fuel burnup takes place. During the first stages of operation, with fresh fuel, the power is concentrated more in the centre of the core, making it a little more difficult for neutrons to escape. Later, as burnup in the centre reduces the central neutron flux, the reactor has more energy in the outer regions and hence more neutrons will escape into the detectors. The gradual production of Pu^{239} in the core as the reactor continues to operate will also complicate the use of neutron counters to determine the power. Nevertheless, neutron counters remain a very valuable prompt indication of general growth and decline.

Water contains oxygen and the cooling water is circulating through the core in intimate contact with the fuel. This gives us the opportunity to measure power based on the interaction of high energy neutrons with O^{16} to form N^{16} by the reaction:

$$O^{16}(n, p)N^{16}$$

It creates the radioactive N^{16}, a nucleus with a half-life of 7 sec. This decays to an excited state in O^{16}, which immediately emits a 6.13 MeV gamma ray as it falls to the ground state of O^{16}. Nature smiles on us for a change since not only is this a convenient half-life, but the gamma ray has an exceptionally high energy, enabling us to distinguish it from the many lower-energy gamma rays buzzing around. It can be detected as the coolant flows out of the reactor, through a shielded measurement station before eventually diminishing due to its 7 sec half-life. The measurement station can be a large 7.5 cm diameter, 7.5 cm thick, sodium iodide scintillation counter, with a discriminator set to preclude all the lower energy gamma background. This method is good for sampling the whole axial and radial extent of the core, but it still suffers from burn-up variations since more neutrons are required to produce a certain power as the enrichment of the fuel decreases. The presence of Pu^{239} also complicates the interpretation since it creates neutrons at a slightly higher energy and the cross-section to produce N^{16} is energy sensitive.

A classic approach to the problem of measuring power is to view the system as a calorimeter. We can use the primary circuit inlet and outlet temperatures, together with the flow rate, to determine the power. It's an excellent idea, but it turns out that the accuracy is not as high as one would like since the mixing of water in the hot leg is somewhat variable and the temperature difference between hot and cold is perhaps only 30°C.

The most accurate way to measure the heat produced in a reactor turns out to be the most complicated. It involves using tables to obtain values for the enthalpy of the steam in the steam generator, enthalpy being the total heat in a system. Enthalpy values, together with flow rates and corrections for various other heat sources, such as that produced by the primary pumps, can give a result accurate to 1%. This method completes the portfolio of options for power measurement, giving operators with different responsibilities the options to view any, or all, of the data.

A quiet day at the office is always better if problems can take care of themselves. Many reactors are designed with this benefit in mind, and they possess natural characteristics referred to as 'named coefficients'. The most common, the ones that you may have heard of, are explained below.

7.11 The Fuel Temperature Coefficient (FTC)

Changes to the temperature of the uranium in the fuel will affect all the nuclear cross-sections, but one cross-section in particular has a marked effect on reactivity. It is the U^{238} neutron capture cross-section, particularly in the resonance region. This process accounts for removing many neutrons from the neutron balance in a PWR. It is not really worrying because it is the beginning of the process for breeding Pu^{239}, a useful fuel in its own right. The fuel at the centre of a pellet is at the heart of the energy-producing region and, in a PWR, the temperature can be as high as 1,000°C. The temperature drops as the heat flows across the temperature gradient to the coolant temperature in the region of 320°C. The temperature effect on cross-sections is known as the Doppler effect which simply reflects the fact that the uranium cross-sections do not just depend on the velocity of the neutron, they depend on the **relative velocity** between the neutron and the nucleus involved. If the uranium happens to be moving towards the neutron as it approaches, then the thermal velocity of the uranium must be added to the velocity of the neutron to obtain their relative velocity. If the uranium is moving away from the neutron, then the relative velocity is reduced. The thermal velocities change with temperature according to the Maxwell distribution, shown earlier in Figure 5.3.

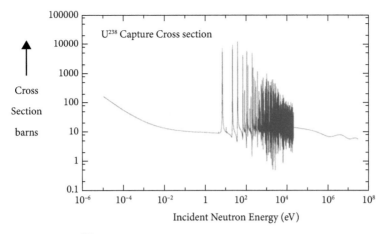

Figure 7.4 U^{238} capture cross-section

Most of the uranium in the fuel is U^{238} and there is a resonance region for neutron capture in U^{238}, illustrated in Figure 7.4.

You should notice the logarithmic scale in multiples of 10, showing how significant these resonances are. The resonances are very narrow and if a neutron has an energy just below the resonance it can be pushed into the resonance if the U^{238} is travelling towards it. The other situation is also effective and if the neutron energy is just above the resonance, then a U^{238} nucleus moving away from it can pull it down into the resonance with a massive increase in cross-section. The overall temperature effect is to broaden the resonances, which leads to more neutron capture by U^{238}. This will remove neutrons and leave a smaller number available for fission in U^{235}. Hence, a rise in fuel temperature reduces reactivity. This effect can be quantified as the **fuel temperature coefficient,** with a typical reduction in reactivity of about 4.0 pcm per °C. Operators can sit back and let a rise in fuel temperature correct itself by reducing the power automatically.

7.12 The Moderator Temperature Coefficient (MTC)

In a PWR the reactor design has a fuel and moderator spacing that result in a slightly undermoderated core. In the case of pure water (without boron) a rise in temperature will cause a change of density, fewer moderating nuclei, and a reduction in reactivity. The effect is larger than the FTC and a typical value for the **moderator temperature coefficient** is −60 pcm/°C. With boron present a reduction in density reduces the effect of the boron, increasing reactivity, so we have a reactivity change in the other direction. In practice the net effect is usually a negative coefficient, but care needs to be taken when the boron concentration is high.

These two coefficients, working together, help to make the standard PWR design a stable reactor under normal operating conditions. I am deliberately avoiding the use of the word safe because it is so often misunderstood. Some of the other reactors may have a positive MTC, notably the advanced gas-cooled reactor (AGR), but since the graphite moderator is separate from the gas coolant, different physics apply and AGR reactors can be operated in a safe manner.

7.13 The Void Coefficient (VC)

The void coefficient addresses the effect of water turning into steam bubbles in the coolant. It has the obvious effect of a reduction in density, similar to the MTC, but needs to be specified as a reduction in reactivity per percentage of void in the coolant. In the case of a PWR it is a quite large, about −100 pcm per % void; a strong negative effect on reactivity. In a PWR the water is not expected to boil but if it does, the reactivity will experience a very rapid fall. It is possible to operate a reactor safely if it has a positive void coefficient and the CANDU reactor is a notable example. The CANDU reactor has rapid, spring-assisted, control rods dropping under gravity, and a gadolinium poison injection system to deal with any safety concerns arising from the positive void coefficient. The Russian reactor at Chernobyl also had a positive void coefficient, which was a contributing factor in the 1986 incident.

7.14 Changes in Steam Demand

An increase in the demand for electricity increases the demand for steam which, in turn, feeds back through the steam generation system to a reduction in the temperature of the cold leg of the coolant. This quickly cools the core and both the FTC and the MTC operate to increase the reactivity and provide the extra power to meet the increase in demand for electricity. It is a reactor following steam demand. This has been achieved without moving the control rods and will result in the reactor operating at a different temperature; not optimal. The operators can return to optimal operating temperatures by either moving the control rods or changing the boron concentration in the coolant.

7.15 Control Room Operations

At the heart of reactor control are the operators in the control room. They are well-trained and have the experience necessary to carry out all the changes that need to take place in the day-to-day operations of a nuclear reactor. You might think that it could, and should, all be done automatically using closed loop control theory, just like setting your room thermostat at home and allowing temperature sensors to feed back into the boiler and keep everything nice and steady, under control. The reality is that nuclear reactors are far too complex to take this simplistic approach. There are simply too many options available when faced with the task of changing the power level of a reactor. The options available are all approved procedures with commercial and operational reasons usually determining which path to take. All reactors have several layers of safety systems overseeing what the operators are doing and any 'mistake' would be recognized and not allowed to happen. To change the power in a reactor, we are going to have to change the reactivity. Let us review some of the various ways we can do this.

The first option that comes to mind is to increase the reactivity by removing control rods, that's what they are there for, but these will alter the axial flux. This is not a safety problem, but it may cause uneven burn-up in the fuel with commercial implications on the eventual life of the fuel.

The second option may well be to change the boric acid loading in the coolant. This affects the reactor in a more even manner and has a similar effect as moving control rods.

The third option is to change the temperature of the reactor by changing some of the externals that will affect the temperature or flow rate of the primary coolant. This will lead to a natural change in reactivity by means of the temperature coefficients in the fuel and coolant.

The other very important consideration, when planning a change of reactivity, is the recent history of the reactor. Poisons, Xe^{135} and Sm^{149} may well have built up and will be changing due to both creation and decay over quite short periods of time. The fuel will have experienced burn up since the last assessment and if burnable poisons are present their effect also needs to be taken into account. The reactor physicists will carefully assess the options and interact with the operators to make it happen. A general principle, that applies to all changes and all reactors, is that all changes need approval for safety purposes. A change requires the new reactivity to be calculated and the target flux to be declared. It is not a case of 'let's make a small change and see what happens'. The effect of any changes must be calculated, approved, and monitored.

When a reactor is being set up to generate electricity from essentially zero power, the coolant temperature and all the externals to the reactor will have been set up in anticipation of heat removal and power production. The control rods specifically designed for start-up will be slowly withdrawn in banks, leaving the other control rods available for final reactivity control. In PWRs there would normally be a high level of boric acid in the coolant, acting as a controllable poison. This level will be reduced as burn-up takes place, probably down to zero as the reactor fuel reaches the end of its life.

During day-to-day operations the operators use their experience, and standard procedures, to implement any changes in reactivity. For a large change in power, the concept of doubling time can be replaced by a start-up rate (SUR), expressed in decades per minute since there are many decades to go from a low neutron population to that present in a high-power reactor.

The change in reactivity may take minutes or hours to achieve. Any change in the boron content in the coolant takes time to travel through the system and become stable and temperature gradients also take time to stabilize. The operators can monitor the change and also predict the point when their target values have been achieved. They may start by changing the boron concentration and allowing it to stabilize. The next step is to start to withdraw control rods to complete the task. They can monitor the effect of this by plotting the reciprocal of the neutron population (i.e. 1/(neutron monitor count rate) against control rod position). When the power is increasing, the reciprocal will fall as the neutron population increases and a graph of control rod position can be used to see, by extrapolation, when the reciprocal rate will hit zero. This is a prediction of the control rod position to create a very high count rate.

Operators may well have to deal with unusual events, faulty instruments, and crisis situations. These should all be covered by standard operating procedures, but experienced operators are there to deal with anything and everything, under the umbrella of the computerized reactor control system (RCS). There were quite a few lessons to be learned from things that went wrong in the first nuclear age and we may well encounter unusual, unexpected, circumstances in the future, but we will deal with them in the manner explained in the next chapter on safety.

8

Safety

8.1 Safety, Risk, and Consequences

Listening to broadcasts during the coronavirus period you were guaranteed to hear the question—is it safe? The public at large, even intelligent people, will ask the same question and I am always tempted to reply with—is it safe to get in a car? The penny drops, of course, and the questioner realizes that there is no simple yes/no answer. The complexity of the subject can be addressed by pointing out that the safety case for a nuclear reactor, which is a requirement for approval to build it, may take more than five years to gain approval and will contain over ten thousand documents to present the case.

Having retired from a career in the nuclear industry, the subject of safety has always been close to my heart, and the subject of many discussions over coffee or lunch, so I make no apology for being generous in the space I devote to this introduction.

My views on safety are related to five keywords, the first two being **risk** and **consequences**. Part of the risk element can often be quantified if data are available, even the risk of an earthquake at a specific strength. Data on the failure rates for a pump should be more reliable if the pump has been operational for a number of years. However, technology is improving all the time and reliability data is usually out of date. Some risks are not so easy to specify but the regulators do rise to the challenge and make responsible estimates in almost impossible circumstances. An earthquake specification may be given as peak ground acceleration of 0.42 g, for example, and a tsunami height as 5.0 metres. In the case of the risk of an exposed core, I have seen declarations of one event per 300 million reactor years quoted by Regulators against another expert's value of one event per 37,000 reactor years by scientists at a University of London college. Perhaps it is a case of a pro-nuclear estimate against an anti-nuclear estimate, so where is the actuarial science?

The **consequences** are usually more complex and depend on each case. A fire risk assessment in a building may not be a worry to the owner because it is a rare event; please forgive the lack of a number. However, if the consequences could be severe, in the case of a storage facility for dangerous chemicals, then the fire officer may insist on an automatic sprinkler system, just in case. The owner may have no choice and must pay the price, which introduces the third term in safety philosophy, **cost**.

Cost is always an element in safety, and I recall the expression—'A responsible degree of safety, achieved at reasonable cost'. This is difficult to argue against, just like 'motherhood' is always good and the words 'responsible' and 'reasonable' are universally accepted as beyond dispute. The trouble is knowing where to draw the line. If

Understanding Nuclear Reactors. Brian Hooton, Oxford University Press. © Brian Hooton (2024).
DOI: 10.1093/oso/9780198902652.003.0008

we look at a security issue the word 'safety' in the same expression could be replaced by 'security', and to illustrate how extra cost may not be justified, let us consider security in a prison. Prisoners will be able to use an exercise yard and there is a risk that a helicopter may fly over, hover, and drop a rope enabling a prisoner to escape. This scenario could be prevented by building a roof over the exercise yard. This would not happen because the consequences of a prisoner escaping are not serious enough to justify the cost. In the nuclear industry the consequences of an 'accident' could be much more severe, and although the industry can justify several additional pumps, to deal with pump failure, since one or more may be faulty or out of service for maintenance, there is a reasonable limit, we don't insist on 20 back-up pumps, and four appears to be an appropriate number in many cases. To have five or six would give extra safety in a theoretical sense, but four is responsible and the cost is reasonable.

Cause is the most enigmatic word to use with safety. There is seldom a single cause, and human failings or shortcomings are often a major factor in the reasons for a calamity. Equipment failures, such as the brakes failing on a car, or an engine stalling on an aircraft, are not difficult to understand but human involvement can be complex. Controlling crowds at major events can go wrong, and it happens regularly. Many fans were locked out at a football stadium and were so desperate to see the match they climbed over a large gate and unbolted it to allow a surge of spectators into the ground; with a crowd of 85,000, the death toll was 33. If the controls can fail it makes sense to assume they will and put the consequences under a microscope. This principle should apply to reactor safety.

8.2 The Regulators

In the nuclear industry, the Regulators have the responsibility for granting licences to build and operate reactors. They define the criteria in a broad sense and, in the UK, the 'Safety Assessment Principles for Nuclear Facilities' 2014, Revision 1 (January 2020) sets out the principles that will govern their approval, or otherwise. They leave the task of setting out the details of a safety case to the organization responsible for building and operating the reactor. It is an interactive process between the operators and the Regulator, taking several years and ending up with a safety case defined in thousands of documents.

8.3 Decay Heat Removal

If a reactor is shut down in unusual circumstances the unavoidable problem of decay heat needs to be dealt with. The critical timeline for dealing with decay heat means that operator intervention cannot be timely enough, so automatic systems have to take the necessary steps.

Consider a loss of electrical power (not a loss of coolant) that shuts down the massive reactor coolant pumps (RCP) in the primary coolant loop. The computerized

reactor protection system (**RPS**) will recognize the problem and initiate a whole set of response activities. It will shut down the reactor and start emergency generators, probably diesel, but other types, such as steam turbines, can also be used to provide emergency power for instruments and pumps. The RCPs cannot be brought back into operation by the emergency generators because they require more power than the emergency generators can provide. Even when RCPs lose their power, they still play a part in dealing with residual heat since they have flywheels and can continue pumping for a minute or so after losing power. This is at a critical stage in the timeline since the decay heat is at its maximum value. After a minute, the decay heat has dropped from 6.5% to 2% but a significant problem remains since it will remain above 1% for an hour, which could be 35 MW for a pressurized water reactor (PWR). Fortunately, the emergency pumps can deal with this amount of heat.

As the emergency generators are starting to operate and the RCPs are slowing down, Mother Nature gives us the added benefit of natural convection to move water through the core. There is an unobstructed path for water to circulate through the core into the steam generator and back into the reactor via the cold leg of the primary circuit. The natural circulation takes time to develop but after 10 to 20 minutes, it is enough in itself to remove all the decay heat. The core should remain covered in this scenario and the reactor protection system has done its job. Operators can now get involved in bringing all systems back under control and normality.

8.4 Loss of Coolant

A **loss of coolant accident** (LOCA) is potentially much more serious than a primary pump failure since the core may become exposed. Nevertheless, the RPS has a way of dealing with it. The RPS will see the pressure drop and the coolant will start to boil, the temperature in the core will rise, but the void coefficient and the fuel coefficient will combine to reduce the reactivity. All these changes will be recognized by the RPS, and the reactor will shut down by the introduction of control rods. So far, so good, but immediate action is needed to replace the water loss and to make sure the fuel is not exposed. For a PWR, the solution is the safety injection system (SI). This is a combination of several pump options, working together, using several water storage tanks, some using a gravity or pressurized feed. They combine to inject water into the reactor via the cold leg according to the overall needs of the system. Overall needs include monitoring water levels and pressures in all the systems external to the reactor. The SI will also monitor and deal with any implications for the release of radioactivity into the environment. It can isolate the reactor building.

This scenario has assumed that it is possible for the safety injection system to achieve its objective. It should do so, but if the cause of the LOCA was a deliberate explosion, by a terrorist, a disgruntled employee, or a mentally deranged individual, then the injection may not be able to prevent a core exposure and a release of radioactivity. In this case, measures to prevent a release of radioactivity into the environment need to come into play to prevent the accident from escalating to an International Nuclear Event Scale Level 6 (Serious Accident) or Level 7 (Major Accident). A well-designed **secondary containment** (SC) should play a big part in preventing the release of radioactive material into the environment outside the reactor building. International Nuclear Event Scales are defined in Section 8.8.

8.5 Passive Safety Measures

Towards the end of the twentieth century, safety measures started to move away from the old traditions of emergency pumps and redundancy. The use of passive measures became an important aspect of overall reactor design and its safety case. A passive system should not need any intervention by an operator, or the use of active equipment to shut down a reactor and enable it to reach a safe condition. It sounds like an ideal solution, but it is not necessarily a complete solution. We have a very limited experience of passive systems in real-life situations, and although the theoretical reliability of gravity is undeniable, can we rely on the passive system to be robust enough to do the job?

Passive systems may be slow and only capable of dealing with a limited amount of heat in the case of a primary coolant incident. There is still a major role for electrically driven equipment, pumps in particular. The use of passive safety features does give a very high assurance that the measures taken to deal with a specific accident situation should not fail. However, it is well to remember that control rods have been known to get lodged in the descent tube whilst falling under gravity. In the case of the passive removal of heat by convection, it is assumed that all the valves are open, not stuck in a closed position blocking the circulation path. There was an accident at the Daiichi plant in Fukushima, Japan, in 2011. Unit 1 claimed to have a passive system, so there was nothing to worry about, but the valves were electrically operated and were closed when electric power was lost. Don't be taken in by claims that passive systems are foolproof.

The subject of safety, especially in reactors, is a specialized topic in its own right, and it is far too complex to deal with here. I will however give a very brief description, and comment, on five nuclear reactor incidents where **uncertainty and confusion** played a part. A much more detailed account can be found in the published proceedings of the investigations into each of them.

8.6 The Windscale Fire

In 1950 the Windscale Pile 1 was operational and ready to produce plutonium for use in weapons. It was fuelled with metallic uranium rods, canned in aluminium, inserted into a graphite moderator. It was a low temperature pile with air cooling by natural convection up through a tall chimney stack, with additional fan-forced air flow, if needed. The original design did not include filters at the top of the chimney stack to remove any radioactive particles, an amazing shortcoming. Sir John Cockcroft intervened, and filters were added during construction, but since the chimneys were already under construction, they had to be added at the top instead of at the bottom. In terms of the key word **cost**, they were considered a waste of money by many and referred to as 'Cockcroft's Folly'. As it turned out they prevented a much bigger discharge of radioactivity into the environment.

The cause of the fire, in 1957, was a natural storage of potential energy in the graphite. This happens because neutrons displace the carbon atoms in the graphite and create a lattice with stored energy. It had been identified and understood and was called **Wigner Energy**. The way to deal with it was deemed to be by raising the temperature of the pile and allowing time for the energy to be slowly released, a form of annealing. Unfortunately, the temperature and time required had not been subjected to extensive tests and during one of the annealing procedures, the energy release caused the graphite to catch fire. It was akin to a charcoal barbeque fire, with ample airflow being sucked up the chimney to make it worse.

Once the fire had been confirmed by a human eye in contact with the red-hot core, somebody had to decide how to put it out, **uncertainty**! The first step was to try and blow it out using the fans. Not a good idea! The next guess was to use water, but since it was uranium, the use of water was questioned, partly because it could act as a moderator, but more likely the decision not to use water was because it could release hydrogen and cause an explosion. The next option was to use carbon dioxide, which happened to be readily available on-site. This did not do the trick, so the water option went ahead in a slow cautious manner. This also failed to extinguish the fire. The final method, which did work, was to cut off all the air supplies to the pile. This did do the trick even, though the burning pile tried to suck air down the chimney.

Once the pile was cool and under control the clean-up started and all the ramifications of the release of iodine and other fission products into the environment were quietly dealt with. This fire did cause a large release of radioactivity into the environment, but it was also a wake-up call with an abrupt awareness of the serious consequences of a nuclear incident. It led to the UK Nuclear Installations Act (1959) and the creation of the Nuclear Installations Inspectorate; now the UK Office for Nuclear Regulation (ONR). It should have created a more international awareness of how to deal with red-hot uranium, and fires in general, but the Brown's Ferry and Chernobyl incidents showed a similar uncertainty about fire abatement. This was the first major incident in the development of nuclear reactors and, although Wigner energy was the technical cause, it could be attributed to the pioneering introduction of nuclear energy into our civilization. We were still very low down on the learning curve, with other lessons to come.

8.7 Brown's Ferry

The Brown's Ferry incident happened in March 1975. We were still climbing up the learning curve in the early days of the nuclear industry, but now under regulations, with safety studies and procedures for dealing with incidents. We had no previous experience of near disasters to fall back on, this was about the first. It would be just one example of the many unanticipated reasons for serious incidents to develop and even more valuable experience on how not to deal with it; we would learn a lot. It showed how factors, other than failure of equipment and automatic safety responses, could lead to a near disaster.

The cause of the incident was a simple fire, which should not have happened, but it did. It was in an area containing most of the electric cables for powering every aspect of the reactor plant. The effect of the fire soon became evident in the control room, where instruments started to misbehave. Lights were flashing, some units were switched off only for them to switch themselves back on. It was utter chaos as equipment failures proliferated. There were two aspects to deal with: the fire and keeping control of the reactor. Both suffered from other issues including poor communication, lack of authority and, in the case of dealing with the offsite consequences, inadequate knowledge of how to deal with a possible release of radioactivity. The incident was not covered by written procedures and the operators themselves used their knowledge and initiative to prevent a major disaster.

The fire took far too long to put out because the plant officers ignored the advice of the professional off-site fire department to use water. They tried powder and CO_2 before accomplishing the task using water. A major concern was activating the 'residual heat removal system' (RHS), which needed to be done manually because of the loss of power. Operators, clad in suits with a breathing system, had difficulty getting to the manual valves before their air supply ran out.

It was almost half an hour after the fire started before the reactor was shut down and operators started to weigh up the consequences. The emergency core cooling system was out of action due to electrical faults, so the two major concerns were the residual heat and preventing the core coolant from falling away to expose the fuel. They knew how to go about it but were without instrumental support since everything was affected by the loss of power. They managed to prevent a serious incident and get the reactor under control after more than a twelve-hour battle.

Telephone communications throughout were slow and problematic because the site had two separate phone systems, with numbers on one system being inaccessible from the other. There were two numbers to use for reporting the fire and sound the alarm, but the complicated phone system confused the operators, who didn't realize there was a fire until 20 minutes after it had started.

There were emergency plans for dealing with any escape of radiation to the outside world but, although the outside agencies knew there was a plan, they didn't have a copy, and later confessed they would not have known what to do if a major release of radiation had occurred. When news of the problem was eventually released it was accompanied by the message—keep it quiet to avoid a panic! This particular point is a genuine concern but not easy to deal with in a written procedure.

8.8 Three Mile Island

The Three Mile Island (TMI) incident occurred after a turbine trip at 4 a.m. on 28 March 1979, and was later deemed to be a level 5 nuclear incident (Accident with Wider Consequences) on the International Nuclear and Radiological Event Scale.

This scale is now often quoted when referring to the seriousness of an incident/accident at a nuclear facility. It was developed in 1990 by the International Atomic Energy Agency (IAEA) and the Nuclear Energy Agency, part of the Organisation for Economic Co-operation and Development (OECD/NEA). The scale is intended to reflect an increase in seriousness by a factor of ten as the scale moves up to the next level. The first three levels are designated as 'Incidents' and the rest as 'Accidents'.

International Nuclear and Radiological Event Scale (INES)

1. Anomaly
2. Incident
3. Serious Incident
4. Accident with Local Consequences
5. Accident with Wider Consequences
6. Serious Accident
7. Major Accident

What happened at TMI can be explained in five significant stages.

The first contribution came eight hours before the turbine trip, from a maintenance operation to clean equipment servicing the secondary loop water. It inadvertently affected an instrument airline which, at about 4 a.m., caused several water servicing pumps to turn off. This in turn caused a turbine trip. This is a good example of the complexity of the engineering side of a nuclear power plant and how non-nuclear problems can escalate into a nuclear incident/accident.

The second stage is the system responding to the turbine trip according to automatic procedures, as expected. The temperature in the Reactor Coolant System (RCS) rose because the steam generators were no longer receiving feed water, and heat transfer to them was reduced. The pressure rose to 155.5 bar and the **pilot-operated relief valve** (PORV) opened, as it should. Steam escaped into the reactor coolant drain tank, which was located in the basement of the containment building. Pressure continued to rise and when it reached 162.4 bar, the high-pressure reactor trip set point, the reactor tripped with control rods falling into the core under gravity. So far so good since all this happened automatically, as expected, only 8 seconds after the turbine trip.

When the feedwater pumps to the steam generators had tripped, three reactor emergency pumps started automatically. Unfortunately, the two emergency feed lines to the steam generators had no effect since both had a block valve closed off. Apparently, this happened two days earlier, as a consequence of system tests. Operating the reactor under these conditions was a violation of the Nuclear Regulatory Commission (NRC) rules. Bending the rules for commercial reasons is a grey area.

When the reactor tripped, secondary system steam valves operated to reduce pressure and temperature in the reactor coolant. The coolant contracted, and since some had been removed by the open PORV the pressure dropped to 152 bar, the point at which the PORV should have reset and closed. The time at this stage was only 15 sec after the original turbine trip. Things can happen really quickly, with not enough thinking time for a human being.

We have now reached the critical stage of the incident. The electric power to keep the PORV open had been cut automatically and it should have closed but it was stuck-open, with the control room indicator saying it was closed. It was not a light saying it was closed, it was the absence of a light saying that it was open. This open valve was the path for a LOCA. There were other consequences and indications that the valve was open that could have been checked to confirm the fact, but the control panel said it was closed and activity in the control room was moving so fast that operators continued to puzzle over the true situation.

The expectations following a LOCA are that the pressure, and the level in the pressurizer, should drop. In this situation, the level in the pressurizer was rising because the valve at the top was open. How can it rise when we have a loss of coolant? With the level in the pressurizer rising, due to the open valve, the incident was not recognized as a LOCA. Emergency core cooling pumps were turned off. Meanwhile, water and steam continued to be removed from the reactor into the external relief tank, which overfilled and discharged into the building sump, raising an alarm at 4.11 a.m. At 4.15 a.m. the relief tank experienced a rupture that released radioactivity into the building itself. Sump pumps were started to pump the radioactivity out of the building but when it was realized that allowing radioactivity to be pumped out of the building might not be a good idea, they were stopped at 4.39 a.m.

The final stage was the consequence of a gradual rise in temperature, caused by residual heat and due to the reactor coolant pumps being switched off. They were cavitating with the steam/water mix. The core became exposed and started to melt.

A secondary problem was the production of additional heat and the generation of hydrogen due to the interaction of steam with the zirconium alloy cladding. This is believed to have caused a small hydrogen explosion.

At 6 a.m., following a shift change, it appears that the PORV problem was rectified but only after some 120,000 litres of coolant had escaped. Contaminated water eventually triggered radiation monitors in the containment building.

At 6.56 a.m., a site area emergency was declared and many outside agencies became involved in the aftermath of the incident at TMI.

8.9 Chernobyl 1986

The Chernobyl story can be over in a flash, literally. There are four significant factors that must be mentioned. The first has nothing to do with technology or equipment;

it is the appalling disregard of safety procedures when increasing the reactor power if Xe^{135} poison is present. Electrical engineers were on-site to carry out a test procedure and had a higher authority than the trained nuclear operators. The electrical engineers forced the nuclear operators, under authoritarian threats, to operate the reactor under strictly forbidden conditions and circumstances. Their efforts to increase the power resulted in a massive and rapid energy excursion as the reactor went critical on prompt neutrons alone.

The problem was exacerbated by the fact that the rate of power increase was enhanced by a positive void coefficient.

The third factor was the design of the control rods. The SCRAM was used but the complex control rod system contained some graphite which caused the reactivity to actually increase for a short period before the boron carbide control poison could take effect. A design that should not have been allowed.

The final important factor was the absence of a substantial secondary containment which allowed the massive release of radioactivity to spread and affect a large portion of Europe. Lamb in the UK could not be slaughtered and eaten.

8.10 Problems in the Fukushima Region of Japan

On 11 March 2011, an earthquake occurred 130 km off the coast of Japan and about 50 minutes later, a massive tsunami surged ashore with catastrophic effects. There were four nuclear plants in the region with 11 reactors susceptible to the effects of the tsunami. The reactors had all been constructed to withstand an earthquake of a certain specified intensity and, regardless of the intensity, all did shut down automatically. The effect of a tsunami had also been taken into account, during the approval stage for licensing. The original specification was based on a historic 3.1 m height tsunami in Chile in 1960. On this basis, the plant had been built 10 m above sea level with seawater pumps 4 m above sea level. The disastrous tsunami of 11 March 2011 turned out to be close to 15 m. An obvious question for both the earthquake and tsunami specifications is, who decides on the adequacy of the specification? Local historic data showed eight tsunamis over the last century with a maximum height of more than 10 m, at the point of origin, out in the ocean. The 2011 tsunami was estimated to have been 23 m at the point of origin. After the Fukushima reactors had been built, the likelihood of a large tsunami continued to be a topic for reassessment, and it remained an ongoing concern, with suggestions to move back-up generators further up the hill, away from a flood risk, but no action was taken.

Two plants, both on the coast and separated by some 11 km, were the Fukushima Daiichi and the Fukushima Daini. The Daini plant was least affected since it had back-up systems protected from floodwater and although the operators had their hands full to deal with the incident, they managed to do it with valuable lessons learned for the future.

The Daiichi plant was not so fortunate and lost all possibility of combating the residual decay heat. There were six reactors on the site, all boiling water reactors (BWRs) but with significant differences in design. Even the earthquake specifications differed, with unit 1 specified as a peak ground acceleration of 0.18 g and units 2 and 3 about three times higher.

Unit 1 had a passive emergency cooling system, but with electric-power-operated valves. These were closed at the time of the tsunami and were not opened automatically after the tsunami due to the loss of power, but could have been opened manually. Units 2 and 3 had steam turbine-driven emergency core cooling systems using steam produced by decay heat. It was a clever idea that failed since they needed electrical power to operate valves and monitoring systems. These were the three units that suffered a meltdown.

The three reactors had been operational from 1971 to 1976. After several expressed concerns about the effects of a tsunami, and placing electrical control systems in the turbine area, which was clearly susceptible to flooding, changes were made. Some emergency cooling pumps were added and placed higher up the hill in the 1990s. There was a policy to share pumps so that they were available for use by any reactor. This required a central control point and, unfortunately, it was placed in the turbine area.

Dedicated, rather than shared, facilities would cost more but the consequences of a failure affecting multiple systems can be disastrous. A reactor may have four separate cooling loops and any shared infrastructure could result in the failure of all four. The failure of shared facilities was relevant in Japan, and at Brown's Ferry where electric cables were susceptible to a single fault escalation.

The IAEA identified a significant conflict of interest in Japan since the Regulator and Ministry for the promotion of nuclear power were not independent.

8.11 Safety Overview

First, a review of the main concerns that we must contend with in nuclear reactors. It matters not what the cause is, be it an earthquake, tsunami, loss of power, sabotage by a terrorist, disgruntled employee, mentally deranged personnel, or any other unlikely reason—we still have to deal with the consequences. The root cause of our concern is the fact that a typical reactor is producing an enormous amount of heat, perhaps 3,500 MW, in a confined space. If the reactor shuts down in response to a SCRAM, there will still be about 227 MW in the residual heat. Dealing with this heat has always been recognized as the main cause for concern. A loss of power to drive the coolant or a loss of coolant itself, a LOCA, must not be allowed to cause damage to the fuel. Related to the coolant problem is the risk of exposing the fuel if the level of the coolant is allowed to fall. Dealing with a failure to prevent a major incident if the core does become exposed then translates into dealing with the consequences, and hoping that the secondary containment does its job.

Many lessons have been learned from all the nuclear incidents since the first reactor CP1 went critical in 1942. Human shortcomings can play a part, as they can in the aircraft industry. Pilot error is often a contributory cause of an accident. Loss of power and equipment failure also contribute to incidents, but redundancy and other measures can be expected to deal with most problems. However, in the case of the unexpected, modern safety standards demand a solution that is foolproof. The search for a foolproof solution can point to using gravity, a force of nature. If convection under gravity is the stated solution let us hope all the valves are open!

The Chernobyl accident, with its massive release of radioactivity all over Europe, reinforced the anti-nuclear opinion that has prevailed in a large fraction of the population since Hiroshima. Germany moved away from nuclear following the democratic mood of the population. In the UK, and probably many other countries, the word nuclear would never appear on the political agenda. It was a negative vote-catcher and with only four years between elections—best avoid it! The fear promoted by Chernobyl is understandable, but it has taken the fear of global warming to renew the political interest in nuclear, even so, the antinuclear lobby remains active. The cost of nuclear power has always been an issue and is obviously compared to the cost of fossil fuels and solar/wind alternatives. Now, in 2023, the steep rise in the cost of fossil fuel and the political need to provide assurance of supply are both moving nuclear power back onto the agenda of politicians.

We tend to say that a major release of radioactivity into the environment won't happen again, but it may, and it is our duty to do everything we can to prevent it. The Regulators, who approve the safety case, and grant a licence for reactors, are now independent from all other interests in most countries, although this was not the case for the Japanese tsunami in 2011. In Europe and North America, regulators and inspectors are given a great responsibility and we believe the reactors to be safe enough. By this, we mean—1 core damage event for every 15 to 20 million reactor years. The statement for the European PR is per 300–350 million years. Unfortunately, these figures are based on somewhat subjective assumptions relating to foreseen scenarios. Unforeseen scenarios, by definition, cannot be taken into account.

We assume that safety standards are improving all the time and not going backwards. This raises the dilemma that if safety standards are better today than they were five years ago, does that make the earlier reactors unsafe? These days they are, after all, designed to operate for up to 60 years, and some hope that small modular reactors will perform for a century. This dilemma exists in the aircraft industry and many other operations that require a safety case. Changes can, and do, take place, the late placement of filters at the top of the chimneys at Windscale as an example. The tsunami specifications were carefully considered before the plants were built but were reconsidered, and thought to be inadequate, after the reactors were operational. Although some changes were made, discussions on the tsunami aspect were still ongoing, with no further action taken, when the tsunami struck.

The need for a secondary containment is being questioned in the safety cases for some of the next generation of reactors. Suggestions that secondary containment is an unnecessary expense in the new era of passive heat removal are appearing on the World Wide Web, on the assumption that the safety of passive systems is absolute.

Secondary containment is there as a contingency to prevent any release of radioactivity into the environment, and should be, in my view, a mandatory feature of all reactors.

8.12 Understanding the Health Hazard of Radiation

Understanding the nature of radiological hazards has always been a problem for the public at large. This is not surprising since it is a fear of something you cannot see, hear, feel, taste or smell. We always have a fear of the unknown and particularly so with anything we don't understand. Telling people to go indoors and shut the windows in the event of a nuclear incident doesn't explain the nature of the problem, it only makes us more afraid—it's like hiding from a ghost. I am going to explain the nature of the dangers of radioactivity in the form of several stories. They are not technical, and when taken together should remove some of the fear that is so common when the word radioactivity is mentioned.

Before I come to the stories, I need to mention a few facts about radioactivity. The effect of radiation on the human body is simply the fact that it has energy that will destroy cells. We know this from cancer treatment and recognize that the dose must be confined to the region of the cancer to prevent it from destroying the good cells that we want to keep. If we get radioactive material in our body, then it will be in intimate contact with our cells and may destroy them. It will continue to do so until we remove the radioactive material from the body by one means or another. There are three types of radiation associated with radioactivity. The first, alpha particles, have difficulty travelling through material, they can get through only 5 cm of air, and will not get through a sheet of paper or aluminium foil. The second type, beta particles, are more penetrating and, although it depends on the energy of the beta, a good rule of thumb is they can get through about 3 mm of aluminium or 30 cm of air. The final type of radiation is the gamma rays, not a particle, more like X-rays or light. These can travel large distances but are less damaging than alpha or beta. Neutrons are also a radiation hazard, but they are only significant near to a reactor that produces neutrons. Radioactivity is almost entirely alpha, beta, and gamma rays.

The secret of minimizing the effect of radiation is to steer clear of it. Keep your distance and remember that if you double the distance, you will more or less quarter the dose. Don't give it any chance to get inside your body. If you come into contact with radioactivity the general advice is to remove your clothing and set it aside, then have a shower to remove any radioactive material from contact with your body. The distance effect means that if you are 1 m away from radioactive material you reduce the dose rate by a factor of 10,000 compared to being 1 cm away. Clearly, swallowing it brings it right into contact with your body cells. This is why, when radioactive dust is in the air, the advice may well be to remain indoors with the windows closed if getting well away from the hazard isn't possible.

The first story comes from the Windscale fire incident in 1957. Radioactive materials were released into the air and one of them was radioactive iodine, I^{131}. This is particularly damaging to humans because it gets concentrated in the thyroid gland and will destroy cells. The iodine release settled on the grassland of the surrounding farms where it was eaten by the cows and resulted in iodine-contaminated milk. The authorities were aware of the danger and the milk was removed from circulation. If it had been consumed, then the treatment was to issue potassium iodide tablets to the public. Swallowing these would increase the iodine level in the body, and nature would discharge it by our natural process of going to the toilet. The tablets would effectively flush the radioactive iodine out of our system.

Nuclear radiation detectors are amazing, they don't miss a thing, and each individual alpha, beta, or gamma creates an electrical signal that can be turned into an audible click, as in a Geiger counter, or displayed by a pointer on a scale. If a counter approaches a radioactive source, it will click away and as it is moved closer and closer, the rate of clicks will increase until it becomes a roar, just as frightening as a lion's roar. When the source is removed the counter keeps clicking away on the natural background, cosmic rays in the main, but also trace amounts of radioactivity in the material surrounding the counter. If the counter is being used to make accurate measurements in a laboratory, then the counter needs to be protected from this background. This is usually done by shielding the counter with lead, on all sides.

When I spent a period working at Los Alamos in the USA, the first thing they did was to send me into a basement for a whole-body scan. This would measure the radioactivity in my body on arrival, and they would repeat the measurement on my departure to show that I had not been contaminated by my work at Los Alamos during my stay. The whole-body scanner looked similar to an MRI scanner in a hospital, but it used radiation detectors shielded, using the best technology available, to minimize the background count in this underground laboratory. Lead was not the best shielding material since it often contained radioactive impurities, so they used steel recovered from a First World War battleship. This was needed to get the background low enough to measure the very low level of radioactivity in a human body; all human bodies are radioactive!

Uranium

Uranium occurs as a metal in the Earth's crust, just like copper, iron, and lead. It is in soil at an average level of three parts per million, and also in seawater, but it has this unique place in our minds as a radioactive hazard, a symbol of the nuclear age. Lead has a density of 11.3 g/cc and uranium is almost twice as heavy as lead at 19.05 g/cc, which makes it useful as a heavy metal. It has been used as a stabilizer in the keel of yachts and has even been used as small counterweights to balance the wings of Boeing aircraft. Uranium salts were also used as a colourful glaze in pottery, well-known for their dazzling green appearance, but uranium's reputation as a radiation hazard has made all these uses a part of the past. The extent of its radioactivity is beyond dispute, and we know that 1 kg of the main isotope, U^{238}, will experience about 12 million disintegrations per sec. This sounds enormous until we realize how little energy comes from a 1 MeV gamma ray. The energy conversion factor means that it would take about two thousand million gammas, at 1 Mev, to raise the temperature of 1 kg of water by 1°C. Even when a Geiger counter is roaring away it may not be dangerous.

I had an occasion to work with a small uranium metal disc, 4 cm diameter and 2 mm thick, and weight of about 48 g. My first problem was to clean it and remove the oxide on the surface. I started to use wire wool but I was warned that uranium is pyrolytic and could catch fire! So, I did clean it with wire wool, but under water. I placed it in a small plastic sample bag, but inadvertently put it into my pocket and ended up finding it still there when I got home. It was a mistake, but nobody seemed worried when I confessed at coffee the next day. What dose might I have received? The sample was quite visible through the plastic bag, which was thick enough to absorb the alphas, although most of the alphas would have deposited their energy inside the sample itself. Just a few, near the surface, would get out of the metal disc and none would get through the plastic bag. 48 g of uranium would yield about 600,000 disintegrations per second, and the radiological effect would be much less than a chest X-ray.

Potassium

Another radioactive element in the earth's crust is potassium. It has one isotope K^{40}, present in a small amount, 0.12%, with a half-life of 1.26×10^9 y. This makes it slightly more radioactive than uranium. It is an essential element in the human body and a well-known component of fertilizer, potash. It is also present in glass, so every time you lift a glass of beer, you are putting a radioactive material to your lips.

Tritium

One of my experiments required the use of a tritium gas target with a Van De Graff accelerator. I was obliged to sign up for the register of tritium users, which put me on the list for weekly urine samples to ensure that I had not ingested a dose. The tritium was supplied in a sealed metal bottle with porous charcoal containing tritium. Heating the container released the tritium through the valve at the top of the container and when we had finished for the day, cooling the container caused the tritium to return to the pores in the charcoal, with no loss to the environment.

It is interesting to compare tritium with uranium as a radiation hazard. Uranium with a half-life of 4.5×10^9 y has a disintegration rate of 1.24×10^7/ s, per kg, but tritium with a half-life of only 12.3 y has a disintegration rate of 3.6×10^{16} /s, per kg. On this basis I was a bit concerned about getting a dose, but an experienced engineer told me, 'No need to worry, if there is a leak it is much lighter than air so it goes up like a rocket'. I stopped worrying, with the realization that if I did ingest a small amount, it would be easy to flush it out of my body by simply drinking gallons of water.

Reprocessing

The final story in this section is about reprocessing. The plants at Windscale in the UK have been operating successfully for many years. The locals supported the industry because it created employment, and they claimed that the only deaths at the plant were those that fell off the scaffolding during the construction. There were several incidents, one being when a storage tank exceeded its critical mass and belched out its contents. A serious spillage but without any serious consequences. There was a lot of opposition to the discharge of 'waste' into the Irish Sea. This was regulated by the National Radiological Protection Board (NRPB), an organization that was stated to be independent of the users of nuclear material. However, the permissible levels of discharge were reduced over the years, more than once, and it was unclear if this was due to better medical knowledge or the ability of the plant to get its discharges down by introducing better technology.

The anti-nuclear lobby made a claim that there was a cluster of leukaemia cases around the Windscale plant and that they were due to radioactive releases into the environment. It is certainly true that leukaemia can be caused by radiation, it is a well-known symptom of 'radiation poisoning'. I had a close friend who died from leukaemia at the age of 40. He had worked on atomic bomb tests in Australia as an employee of the UKAEA, not the military, but was, however, treated in a military hospital. As is often the case, his death was not definitely linked to his time in Australia since leukaemia can be due to causes other than radiation. The enquiry into leukaemia clusters went on for years but my recollections are that a statistically significant cluster was identified near a lead-smelting factory in Yorkshire and another way back in the 1930s, way before fission. The findings of the enquiry were inconclusive.

When you call for the services of health physics, they will often arrive with a hand-held monitor. They will carry out a detailed survey of the area and then tell you how long you can continue to work at the prevalent level of radiation. They may well tell you what the radiation level is and, if you are not a health physicist yourself, you will have difficulty understanding the language. Terms and units for levels of radiation have undergone many changes. The early unit, the Roentgen, was defined in 1928 and related to the amount of electric charge generated by the dose. Several other units, the Rad and the Rem, appeared but the current international units are the **gray (Gy)** and the **sievert (Sv).** The gray is not difficult for a physicist to understand: it is the amount of energy, in Joules, absorbed by 1 kg of matter. This recognizes the fact that if the energy is deposited over a larger mass, then its effect is diluted and becomes a diminished health effect. Unfortunately, this does not give a direct indication of the likely health effects. The sievert is the amount in Gy, modified to take into account the type of radiation and the likely effect on the various organs of the body, which takes us into the very complicated realm of radiation medicine. There are so many names to contend with, external dose, operational dose, equivalent dose, and dose equivalent. The USA have some of their own names for units of dose, and even the rationalization of the names suggested by the 3rd Symposium on Radiological Protection as recently as 2015 has been taking time to come into common agreed use. In practice, the UK Radiation Protection Act limits the value for occupationally exposed persons to 20 millisieverts per calendar year, and the exposure to members of the public to just 1 millisievert per calendar year.

The message in these stories is that, if you work in the nuclear industry, make sure you don't get radioactive material on your skin or ingest it. Follow the safety procedures for your facility to the letter. If you are supposed to check your hands in a radiation monitor before leaving a restricted area, make sure you do, and if in any doubt, contact Health Physics. Members of the public should not have to worry about radiation hazards, but if there is a radiation leak into the environment, take note of public announcements. One final piece of advice if you want to live to become an old-age pensioner: don't drink and drive.

9

The Nuclear Fuel Cycle

9.1 The Nuclear Fuel Cycle Definition

The nuclear fuel cycle diagram, shown in Figure 9.1, is a useful way of understanding all aspects of the commercial production of electricity from fissile material. The stages necessary to produce the fresh fuel are referred to as the front-end of the cycle and the stages after the spent fuel has been discharged from the reactor are referred to as the back-end of the cycle. If the spent fuel is simply stored the cycle is called an open cycle, but if fresh fissile material is produced by reprocessing and reintroduced into the front-end of the cycle, it is referred to as a closed cycle. Any new reactor concept needs a full evaluation of the impact of its particular fuel cycle on the pros and cons of using the design for commercial purposes. There is the development cost of any special fuel manufacturing plant and the question of how to store or reprocess the spent fuel with due regard to the question of closing the fuel cycle and re-using some of the reprocessed fissile material. Many of the fuel cycle issues could require extensive

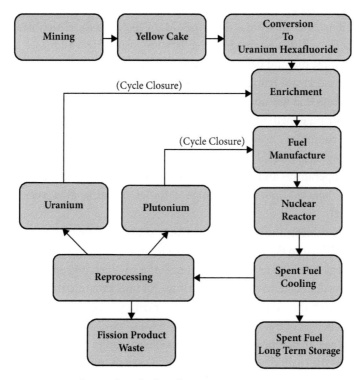

Figure 9.1 The nuclear fuel cycle

Understanding Nuclear Reactors. Brian Hooton, Oxford University Press. © Brian Hooton (2024).
DOI: 10.1093/oso/9780198902652.003.0009

design and experimental verification outside the design of the reactor itself. All these questions relate to the final cost of producing electricity.

9.2 Mining

The main sources of uranium ore are in Kazakhstan, Canada, and Australia, with significant supplies in China, Russia, the USA, and elsewhere. Underground and open-cast mining, similar to coal mining, is used to extract the ore, which is then crushed, leached, and processed to recover the uranium. An in-situ leaching process is also common, where a leaching solution is pumped underground, and the dissolved uranium is pumped back to the surface. Both methods conclude with chemistry to form a yellow or dark brown oxide, U_3O_8, not yet the black UO_2 usually used in fuel manufacture. The U_3O_8 product is commonly known as yellow cake. There is also a small amount of uranium in seawater, and Japan carried out experiments to extract it in commercial quantities, but economics led to the abandonment of this approach. Figure 9.2 shows the components of a typical uranium mill to produce yellow cake.

9.3 Enrichment

The next stage in the cycle is normally enrichment in U^{235} content, but this step may be bypassed if natural uranium is the target. Many methods for enrichment have been used ranging from diffusion, magnetic separation, centrifuge, and laser enrichment, with the centrifuge method being the main workhorse for enrichment at the moment. The laser method is still being developed and should have a commercial advantage, but there are fears that it is not as non-proliferation friendly as the centrifuge method.

The gas centrifuge method requires uranium in the gaseous form of UF_6, uranium hexafluoride, commonly referred to as HEX. It has an unusual phase diagram with all three phases, solid, liquid, and gas, being seen at some time during its life under processing temperatures and pressures. The phase diagram is shown in Figure 9.3.

The gas is injected into a centrifuge cylinder, rotating at a very high speed (see Figure 9.4).

The rotation subjects the molecules to a large centrifugal force which preferentially separates the U^{235} from the U^{238}, with the heavier isotope being taken out of each stage at the outer regions of the centrifuge; this makes a small change in the isotopic composition. Hex is then taken out of the centrifuge for injection into the next stage, which will further enrich the output. The series of centrifuges used to enrich the uranium, bit by bit, is called a centrifuge cascade.

The centrifuge enrichment process has nuclear weapon implications and almost every aspect of the plant design was treated as classified information, particularly the speed of the centrifuge itself. Unfortunately, the plants were open to visitors, including foreign nationals working as International Atomic Energy Agency (IAEA) inspectors,

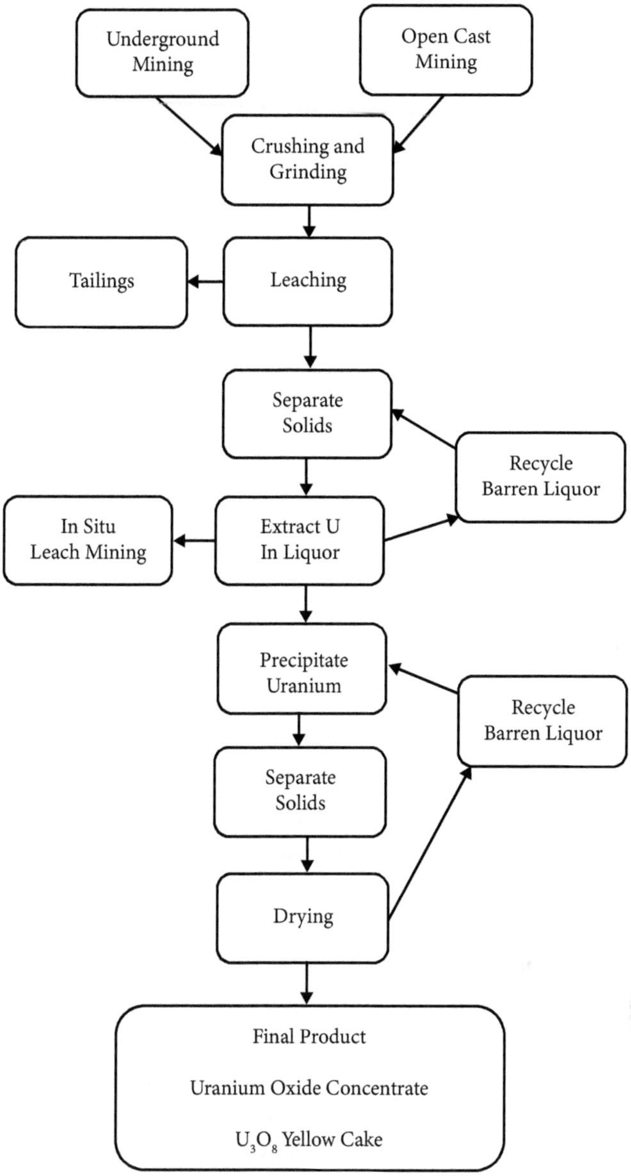

Figure 9.2 Uranium mining to produce yellow cake

and if they had any ear for music at all they would immediately recognize the hum of the centrifuge as close to concert A!

The UF_6 output is stored, under pressure, in a transportable tank for delivery to the next stage in the cycle.

9.4 Fuel Fabrication

The most common form of fuel in current reactors is UO_2, a black powder formed from UF_6 by various chemical flow sheets. It is compressed into pellets and sintered

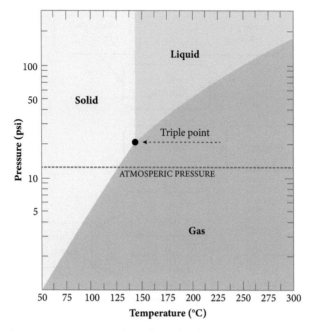

Figure 9.3 Uranium hexafluoride phase diagram

with a binder, a conventional powder metallurgy process. The result is UO_2 fuel pellets shaped as small cylinders about 1 cm diameter and 1 cm long. These are inserted into fuel pins which are then amalgamated into groups to form a fuel assembly. Quality control is essential and since the pellets with a different enrichment will look the same, checks are required to ensure that a pin does not contain a 'rouge pellet', one with the wrong enrichment. A rouge pellet could result in a hot spot and pin failure, with a release of radioactivity into the reactor coolant.

The pin cladding material is chosen for engineering, corrosion, heat, and nuclear properties. Some of the early reactors used a magnesium-aluminium alloy, MAGNOX, but it was abandoned for stainless steel or zircalloy because the alloy had problems during reprocessing and long-term storage under water. This is a good example of how the requirements of the fuel cycle as a whole need to be taken into account when selecting a cladding material.

Although UO_2 is the most common form of fuel, uranium in any form is a candidate for initiating a chain reaction. Uranium metal was used in the early reactors and high-temperature forms such as uranium carbide have been used in several reactors. The use of soluble forms, introduced as part of the coolant cycle, is a possibility, and composite forms of pebbles abound.

9.5 Spent Fuel Management

The lifetime of fuel in most reactors is determined by the build-up of unwelcome poisons. The notable poison Xe^{135} decays naturally and does not present a long-term

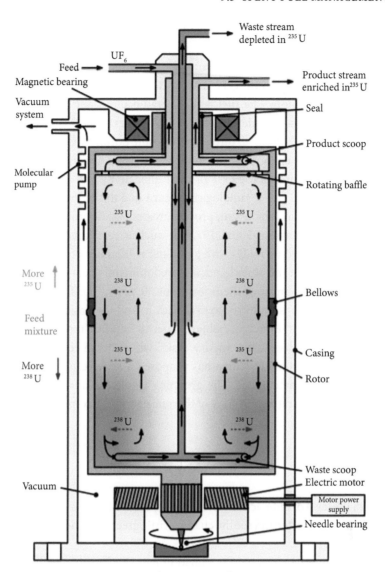

Figure 9.4 A uranium enrichment centrifuge

Source: Inductiveload

problem, but Sm^{149} is stable and with a neutron absorption cross-section of about 40,000 barns, the large negative reactivity eventually results in the fuel being declared 'spent'. The first destination of spent fuel must deal with the high level of radioactivity in the fuel and the fact that it is thermally hot and will require cooling. The solution to these problems is the spent fuel storage pond, where they are stored underwater. This provides protection to the environment from any radioactive leak and keeps the fuel cooled to the temperature of the pond. The time spent in the pond is going to be many years as the radioactivity slowly dies away. It is correctly said that some activity remains for thousands of years. This passive storage of spent fuel will continue until the problems of handling a highly active item become manageable. It can then be moved to another form of storage or moved for reprocessing.

9.6 Spent Fuel Ponds

Spent fuel storage ponds have the appearance of a swimming pool, but they are not for swimming! They are typically 12 m deep with structures at the bottom to fix fuel assemblies in position. The fuel is highly radioactive, and also thermally hot, so it needs cooling. This is achieved by heat exchangers, with the pool temperature kept below 50°C. The bottom fixtures may contain boron poisons to ensure sub-criticality. The thermal energy and the radioactivity will decay naturally over time and after some 30 to 40 years, 99.9% of the radioactivity will have decayed away. Nevertheless, the fuel is still a radiological hazard due to the long lifetime radioactive products that remain. The fission process creates fission products in two groups cantered around the masses 95 and 137, but nuclear reactions with uranium can result in many long-lived actinides with alpha decay lifetimes of thousands of years. The fission products, on the other hand, are predominantly beta decay isotopes with much shorter lifetimes, most of them less than 30 years. The fuel may stay in the pond for decades, there is no hurry to deal with it, and the passage of time reduces the activity and makes the fuel assemblies easier to handle. In some cases, the fuel may be transferred to dry storage in 'dry casks' where it can remain, with a lower management cost, indefinitely. China has proposed a scheme to amalgamate spent fuel assemblies in a reactor pool. The grouping would be capable of sustaining a chain reaction and generating 200 MW of thermal energy for community heating and desalination.

9.7 Cherenkov Radiation

Even though the level of radioactivity in a storage pool is high, it is still possible to stare into the depths of the water and see the characteristic greenish-blue glow of Cherenkov radiation.

Pavel Cherenkov first observed the radiation named after him in 1934, but Marie Curie had noted similar bluish light way back in 1910. The glow is due to photons emitted by a charged particle moving at a speed greater than the velocity of light. Impossible, you might say, since nothing can travel faster than light, but the velocity of light in water is about 30% lower than its velocity in a vacuum and an electron travelling in water can exceed the phase velocity (the speed of propagation of a wave front in a medium) of light. The theoretical explanation came in 1937 by Igor Tamm and Ilya Frank, who shared the 1958 Nobel prize with Pavel Cherenkov.

9.8 Reprocessing

The first comment on reprocessing is that some countries support the USA initiative not to carry out reprocessing because it can result in the production of weapons-grade plutonium. This policy was announced by the US President Carter in 1977.

The influence of the USA on this aspect of the nuclear fuel cycle resulted in many nations, including the UK, moving away from reprocessing in favour of long-term storage of spent fuel. This does however leave the option to reprocess in the future as uranium supplies become scarce, and if the current optimism for breeder reactors takes off, then reprocessing may well be back on the agenda, regardless of US pressure. India, for example, is well on the way to using the thorium breeding cycle since it has its own plentiful supplies of thorium.

The UK Thermal Oxide Reprocessing Plant (THORP) is a typical example of the reprocessing option for dealing with spent fuel. The first step, the head-end, entails feeding the fuel horizontally into a shear cutting device. This consists of a very massive guillotine with a cutting edge and energy sufficient to drop down and chop off a chunk of the fuel assembly. It operates in a similar manner to the guillotine used in the French revolution. The analogy goes further since the shear deposits a chunk of the fuel assembly into a basket. The name head-end did not take its name from the French revolution! The stainless-steel basket contains drainage holes and is sitting in a container filled with nitric acid, called the dissolver. Most of the elements in the fuel will dissolve to form a nitrate which can go forward as a liquid batch to the next stage. Once the dissolution is complete the basket containing the spent hulls can be removed for monitoring and storage. They will contain residual amounts of fuel trapped by the crimping action of the shear. The dissolver batch will be sampled and analysed for nuclear accountancy and quality control purposes.

The next stage implements the plutonium-uranium extraction (PUREX) chemical flow sheet for the separation into three streams: a uranium stream, a plutonium stream, and an 'everything else' stream. The flow sheet is presented in Figure 9.5.

The 'everything else' stream, containing all the fission products and several actinides, is often referred to as the 'high-level waste' stream.

The separation technology required must be as free from human intervention as possible because of the high radiation environment. The method at THORP uses mixer-settler devices, illustrated in Figure 9.6.

The product and solvents are mixed using electric stirrers in the mixing stage, and reactions take place as the chemicals flow along a trough, like a slow-moving stream. As they travel, the lighter component rises to the top and the denser component sinks to the bottom, so that, at the end of their travel, they have been separated by gravity. The light component overflows a high-level weir and the low component flows over a second weir, set at a lower level, thereby separating the reaction products. With more sophistication and care than I can describe here, the final products are extremely pure, with very little cross-contamination between the three streams. The uranium and plutonium are stored as oxide for feed back into the fuel cycle and the 'high-level waste' (HLW), moves to a finishing plant. The final stage for the HLW is a reduction in volume before storage in a double-walled stainless-steel containment tank. This syrup-like liquid is thermally hot and needs to be cooled with a heat exchanger.

The HLW storage tank is not considered to be the ultimate form of storage since it is in a liquid form and any disruption of the tank would lead to an unacceptable

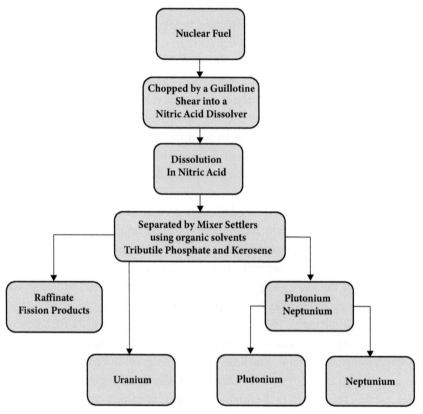

Figure 9.5 The PUREX process

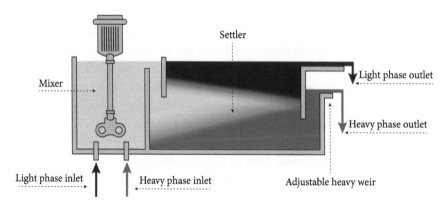

Figure 9.6 A mixer-settler

contamination of the environment. The final resting place for this very dangerous material is discussed in Section 9.9 on nuclear waste.

The uranium output after reprocessing has a different isotopic composition because it has been in a reactor. It will contain a small amount of U^{232} coming from the decay of Pu^{236} after a lengthy time in a storage pond. This is not a problem since it captures a neutron and becomes fissile U^{233}. The other un-natural isotope is U^{236}, which presents a minor problem since it is a neutron absorber to the extent that it requires compensation, by using an increased U^{235} enrichment.

The plutonium output has a quite variable isotopic composition, depending on the fuel's burn-up history, but it is pure enough for direct use, which is usually as a component of MOX, a mixed oxide of UO_2 and PuO_2.

9.9 Nuclear Waste

The nuclear fuel cycle produces 'nuclear waste' at every stage, but some of it has not been near a reactor and can be considered as in the 'low-level waste' (LLW) category. This includes overshoes worn in areas where radioactive spillage might occur, rubber gloves, and paper tissues, so it is not surprising that quite a lot of it is not contaminated. The other feature of this category is its large volume which is often reduced by compaction or even incineration. In the very early days of the nuclear industry, some of the LLW was mixed with concrete, put into barrels, and dumped into the middle of the Atlantic Ocean, a practice frowned upon by many and eventually discontinued. Scientific arguments were presented to support the dumping on the grounds that even if small amounts of this very LLW leached out of the barrels at the bottom of the ocean, the radioactivity would be so dilute it would be less than the cosmic ray background. Other arguments referred to the concentration effect of mercury poison in fish which had become a problem in the Pacific and in the end the uncertainty factor won. Any form of disposal that was not retrievable has the disadvantage of not being able to overcome the uncertainty factor. LLW is now checked by instruments to confirm the content and put into a form for land storage in the most economical manner.

At the other end of the spectrum is 'high-level waste' (HLW). Most of this category is identified as the output of fission products and actinides from reprocessing, everything apart from the uranium and plutonium. This really is nasty stuff, a thick concentrated liquid with so much internal radioactive energy it would bring itself to the boil. It is stored in a very large stainless-steel tank with water cooling pipes to keep the temperature down. It could remain there forever under normal circumstances but would remain a hazard in the event of an earthquake, a terrorist attack, an aircraft impact, or a missile. The long-term solution is to convert it into an inert solid, and this is done by mixing it with sand or another stone-based product and then heating it until it becomes similar to a solid block of glass. The technical term is **vitrification.** I like this solution because I think of it as a Coca-Cola bottle, if you drop it on a stone floor it will break, but so what—it will not contaminate the environment with radioactivity, and it isn't going to go anywhere. The long-term fate of vitrified HLW is expected to be in a geological depositary, where it can sit for thousands of years. It is intended that the vitrified blocks will be sealed in a stainless-steel container to prevent the escape of gases, and these will be placed underground in a stable geological formation. A simplified design concept is to drill a vertical shaft and then drill horizontal shafts, like the spokes of a wheel, at several deep underground levels. Once a spoke has been filled it would be 'bricked up' to seal it and should be capable of existing for thousands of years. However, it would need to be monitored and, in the event of an unforeseen problem, then an intervention would be required. This would not be a problem because, thank goodness, it is retrievable. None of these storage facilities are yet up and running. Some long-term storage facilities are being designed with the claim that they need not be retrievable.

Spent nuclear fuel under storage in a pond can be considered as HLW, awaiting treatment, but it contains valuable fissile material, so the word waste is misleading. Some spent fuel is to be stored in dry facilities and Finland has a geological repository for fuel in copper containers, due to receive waste in 2024. It is claimed it should be safe for 100,000 years.

The delay in making a political decision on how to dispose of waste is said to be because the technology is changing all the time and we can afford to wait. This tends to make the indecision ongoing forever, and perhaps we will eventually send the waste into space, directly into the sun. This may seem unwise, but we are already sending human beings into space as tourists, and if a cargo did come back to Earth, the packaging could be designed to enable prompt retrieval and avoid any contamination.

Intermediate-level waste (ILW) is, as its name suggests, everything between the two extremes. It can include material removed from any plant in the fuel cycle and from the decommissioning of reactors, a very variable category. ILW is managed according to its content, and this is always determined by advanced instrumentation, capable of measuring gamma rays and neutrons coming out of each waste container. This is a speciality in terms of the expertise required. The contents are examined slice by slice, down the container, as it is rotated, with the intention of identifying and confirming the contents. It is a process similar to an MRI scan in a hospital. Once the contents have been verified, the container can be sentenced to the appropriate packaging and channel for storage.

The nuclear fuel cycle is evolving as new proposals for fourth-generation reactors come to fruition. New non-proliferation reprocessing initiatives will be needed to implement some of the new reactors and new fuel technology will replace the familiar UO_2 pellets. Some of these changes are discussed in Chapter 11.

10

International Treaties and Obligations

10.1 Euratom

Two treaties were signed on 25th March 1957 in Rome, one establishing the European Economic Community (EEC), now called the European Union (EU), and the other creating **Euratom**, the European Atomic Energy Community. This was a time when there was a general belief that oil and gas supplies would run out by the end of the millennium. North Sea oil and many other reserves were yet to be discovered, as was the technology of fracking. Nuclear power was reckoned to be the salvation of the world's economic progress. This vision created a fear that uranium might soon become in short supply and would create a battle, at the national level, to ensure the security of supplies. Euratom addressed this problem in Europe by creating a European Atomic Energy Community to guarantee a secure supply for all its members. The treaty embraces all the commercial aspects associated with nuclear materials. It covers, amongst other things:

Assurance of supplies of ores and nuclear fuels.
Supervision to give assurance that nuclear materials are not diverted from their intended purpose.
To establish uniform safety standards.
To promote research.
To ensure wide nuclear commercial markets.
To exercise the right of ownership conferred upon it with respect to fissile materials.

This last item is covered in Chapter 8, Article 86. It states, 'Special fissile materials shall be the property of the Community'. The word 'property' is to ensure 'Equal access to sources of supply'. Special fissile materials (SFM) are defined as enriched uranium and plutonium and the treaty requires, in theory at least, all members to have the consent of Euratom to use SFM in reactors. This is just one of many aspects of Euratom since the treaty covers aspects that go beyond the use of SFM. A notable example is the support of nuclear fusion research with the Joint European Torus (JET), and the International Thermonuclear Experimental Reactor (ITER) tokamak, an international collaboration devoted to the development of fusion as a source suitable for the generation of electricity.

Understanding Nuclear Reactors. Brian Hooton, Oxford University Press. © Brian Hooton (2024).
DOI: 10.1093/oso/9780198902652.003.0010

10.2 Treaty on the Non-Proliferation of Nuclear Weapons (NPT)

The Treaty on the Non-Proliferation of Nuclear Weapons (commonly known as the NPT) was opened for signature in 1968 with Russia, the USA, and the UK as its founding members. Its purpose is to:

> Prevent the spread of nuclear weapons and weapons technology, to promote cooperation in the peaceful uses of nuclear energy, and to further the goal of achieving nuclear disarmament and general and complete disarmament.

When it came into force in 1970 it had 40 members, in addition to its three depositary states. France did not sign at that time but said that, although it had not signed, 'it would behave as though it had signed'!—a political nuance. They eventually joined in 1992. The treaty was reviewed in 1995 and is now a permanent treaty. The agency responsible for monitoring compliance with the treaty is the IAEA.

10.3 The International Atomic Energy Agency (IAEA)

The IAEA is independent of the United Nations by means of its own international treaty, but it is strongly linked and identified with the UN since it reports to the UN General Assembly and the Security Council. It is now generally considered to be a UN organization. It has a very wide remit, one of which is 'Safeguards', nothing to do with safety in the conventional sense, but everything to do with—'preventing diversion of nuclear energy from peaceful uses to nuclear weapons or other nuclear explosive devices' NPT Art. III.

10.4 Nuclear Safeguards

Nuclear Safeguards is the system for monitoring compliance with the two treaties, the Euratom Treaty and the NPT. Each treaty has its own Safeguards approach but there is a significant overlap of interests and, to minimize duplication, there is a degree of collaboration between the two organizations, the Directorate of Euratom Safeguards and the IAEA. The main difference is an academic one, that the IAEA have a special requirement to identify noncompliance in a timely manner. This means, in a general sense, that the IAEA inspection activities should reveal a risk of a bomb being produced before the culprit has the time to make a bomb.

The policing of both treaties includes inspections, audits, and compliance with regulations on the management of nuclear materials.

Nuclear material accountancy is done by bookkeeping and stocktaking, to give assurance that nuclear materials have not been diverted from their declared use. Each type of material has a separate account. The main accounts are:

Depleted uranium
Natural uranium
Low enriched uranium LEU U^{235} below 20%
High enriched uranium HEU U^{235} 20% or more
Plutonium (regardless of isotopic composition)

In commercial nuclear operations, any quantity that is moved from A to B or transformed in its composition, like uranium fluoride to uranium oxide, for example, must be recorded and the relevant account changed. Safeguards can audit these accounts just like financial auditors deal with financial accounts. Records can be verified by a statistical sampling technique and any selected items can be verified by measurement. If a compound of uranium is weighed, the uranium content must be calculated by using the stoichiometric composition, with additional corrections for water vapour or any other impurities. It is just the uranium content that is used in the account.

Measurement is not trivial in the case of nuclear accounts. In the case of plutonium, the material will often be in a container but may be in a process stream like a glove box, or in liquid form in a reprocessing plant, or in a spent fuel storage pond. Weighing a container will give a value for its contents if the weight of the container is known, but what if the plutonium has been substituted with lead, a cover-up? The plutonium in the container may not be dry and, assuming it is PuO_2, a correction will have to be applied for its chemical composition. Not the same as counting cash.

Technology gives us several options for arriving at a figure for the actual plutonium in its container. Safeguards inspectors will bring their own specialized non-invasive detection systems to carry out on-site measurements. The gamma rays emitted by plutonium in a container can be identified and used to determine both the quantity and the isotopic composition of the plutonium. Plutonium releases neutrons by spontaneous fission, so neutron detectors can also be used to measure the quantity, but since the neutron emission varies from isotope to isotope, the isotopic composition needs to be known for this method to achieve an accurate result. The neutron method also requires a correction for neutrons from other sources, such as neutrons generated by alphas, coming from the plutonium, and interacting with oxygen by means of an (α, n) reaction.

Plutonium containers will be warm to the touch, because of the heat generated by the radioactive decay, which means that, if the isotopic composition has been measured by gamma ray measurements, a heat calorimeter can be used to arrive at a value of the total plutonium mass.

These sophisticated measurement systems can be used for uranium and plutonium, but the time taken to do them for every item would be far greater than the time available for a busy inspection team. Fortunately, most of the inventory does not change from one inspection visit to the next, so measurements are not necessary on every visit if the containers have been sealed by the inspectors with a tamper-proof seal. This enables the inspector to simply check the seal and confirm the amount of the item, and then tick it off in the book account.

As well as accountancy, the inspectors need to keep an eye out for any undeclared activity. They will often have cameras at key places to record activities for analysis later, back at the office. In the case of an enrichment facility, they can confirm the percentage of U^{235} by gamma ray analysis, and may take samples from pipes in the plant for analysis back in Vienna to give more accurate results.

10.5 Obligations

The two treaties with a broad international flavour, the Euratom Treaty and the Treaty on the Non-Proliferation of Nuclear Weapons (NPT), do not complete the requirements for overall legal compliance. There are several bilateral agreements between nations for the supply of nuclear materials. They are often under the heading of 'Nuclear Cooperation Agreements', covering many aspects in the nuclear field in addition to supply. They often demand additional obligations on the use of the nuclear material. Looking at the UK–Australia Agreement as an example, it states the conditions that Australia imposes when it supplies nuclear material, to ensure it does not get used in weapons. When the UK was in the EU, Euratom would need to be involved, and it would be a tripartite agreement, but since BREXIT new cooperation agreements with Australia have come into force. The original UK–Australia Agreement stated that the nuclear material could not be used in weapons, or for any other 'military purpose'. This would include the use in nuclear submarines but, with mind-stretching interpretation, could mean that the electricity produced in a nuclear reactor could not be used in a military base or the medical isotopes produced in a reactor could not be used in a military hospital; both of these being a military purpose. The new UK–Australia Cooperation Agreement, 2021, has been amended and provides a definition of 'military purpose' that now allows indirect uses in military bases and hospitals.

The other restriction that may be more relevant in the future is a restriction on reprocessing. This became a policy with US supplies because it is a pathway to separated plutonium, with obvious weapons implications.

To legally confirm that obligations are being met, a system of 'obligation pool accountancy' has been adopted. There is an accepted principle of equivalence that one atom of uranium is indistinguishable from the next. This is somewhat like saying that one dollar bill is the same as any other. Each entry in the records for nuclear material will carry a label (e.g. S for Australia) stating that this material is deemed to be under Australian obligations. The letter N is used for non-obligated material, meaning it could, if there were no other reasons not to, be transferred from the civil account in the UK into the military cycle. This system of obligation pool accountancy enables each nation to monitor compliance with the requirements of its agreement.

11

The Future of Fission Reactors

11.1 The Alternatives to Fossil Fuel

Global warming has been the catalyst for the development of new paths to energy production, without the formation of CO_2. This chapter examines the role of nuclear energy in the twenty-first century. It concentrates on the expected development of reactors as Generation IV technology emerges but recognizes that solar PV and wind turbines are providing much cheaper electricity as their technology also advances. Enormous solar farms are being built in remote deserts with energy storage and hydrogen production helping them to become viable sources of electrical power. Wind turbines in offshore locations, or built on artificial islands, complete the third main option for producing electricity without the release of CO_2. Industry is always looking for the cheapest source of energy, but political pressure to provide an assurance of supply, 24/7, plays a significant part in making decisions on which of these three alternatives should be supported. In most cases it is considered to be better, and wiser, to sit on a three-legged stool giving a better balance to the stability of supply. Section 11.13, on the economics and politics of electricity, explores the basis for making investment decisions.

11.2 Generation IV Technology

The introduction of Generation IV technologies will take time. Most of them are not original, but they push the realm of operation well beyond the capabilities of Generation III. Demonstration plants will be needed, and problems will be encountered without doubt, so we can expect some of them to be abandoned for commercial reasons. Nevertheless, we can assume that the technical difficulties will be overcome in due course, and I will concentrate on the theoretical advantages of all the proposed developments. Most of the so-called new innovations have been tried before, molten salt reactors, gas coolants, metal coolants, breeders, and high-temperature operations. The main criteria for making a choice for any new reactor are safety, reliability, and cost, but the technology is emerging with plenty of new ideas and new challenges that will engage the ingenuity of today's scientists for many years. An unfortunate delay on the path to net zero is it still takes years to obtain a design acceptance confirmation (DAC) or a statement of design acceptability (SoDA). This can only be given after the third step of the generic design assessment (GDA) has been completed. In the UK, in June 2023, the Rolls Royce SMR has just completed Step 1, with Step 2 expected to

Understanding Nuclear Reactors. Brian Hooton, Oxford University Press. © Brian Hooton (2024).
DOI: 10.1093/oso/9780198902652.003.0011

take 16 months before the last Step 3 can begin. I did say unfortunate, but these safety assessments are essential since we must never have another Chernobyl.

The changes will be significant, with a surprising move away from thermal neutrons to fast reactors in many of the Generation IV concepts. A move away from water as a coolant is perhaps not such a surprise since the development of higher-temperature operations has many benefits. New reprocessing technology will emerge, but it may be several decades before commercial plants come into operation. The term 'integral reactor' has started to appear with several slightly different meanings. It can mean that all the inner workings of a reactor, including steam generation, primary cooling, and any emergency cooling are all contained within a single pressure vessel. The other interpretation is that integral means that the other parts of the fuel cycle, reprocessing, and even fuel manufacture and waste storage, all take place on the same reactor site. This is desirable if the reactor needs its own closed fuel cycle.

The new developments can be presented in terms of the generic changes that are taking place without too much reference to a particular model. Safety is still top of the list in all cases, so let us examine the generic changes one at a time.

11.3 The Move to Higher Temperatures

The move to achieve higher temperatures means a move away from water as a coolant. It will lead to higher thermal efficiency, with the possibility of gas turbines giving an even higher efficiency because they use the Brayton Cycle. Several coolants are possible: molten salts, liquid metal, and gas. New types of fuel, not the standard UO_2 pellets, are proposed to enable reliable performance at higher temperatures. Higher temperatures are also attractive since it takes us into the realm of process heat, a fascinating area with many possibilities, including the production of hydrogen. However, before we get too excited, let us examine the difficulties ahead. The pressurized water reactor (PWR) operates at about 300°C and the move to higher temperatures in the region of 600°C is not too difficult, we have already been this high, but this is not really high enough for efficient process heat, we need to go beyond 800°C, and even 1,000°C is not always enough. The journey to higher temperatures will have to be gradual and when you realize that a red-hot poker is just 800°C, temperatures in this region will entail reactors glowing bright red in the dark. Providing the structural materials to operate at these temperatures, continuously for about 40 years, will be quite a challenge.

11.4 The Move to Fast Reactors

Since we are not yet building fast reactors, this suggestion may come as a surprise to many. Conventional logic told us that we need to get down to thermal neutron energies to take advantage of the high-fission cross-section in fissile materials. It was all about neutron economy, to avoid capture, and to use an efficient moderator. Fortunately, when breeder reactors were examined in the twentieth century, and it was shown that they would need to operate as fast reactors with neutron energies in the region of 1 MeV, we revealed design features in fast reactors that have some added benefits.

Fast reactors will only work with a higher enriched fuel, over 6% U^{235} or Pu^{239}, being the starting point. They will be more compact with a higher energy density, about 1 kW /cm^3, requiring efficient heat removal and operating at a higher temperature than thermal reactors. Liquid metals were initially proposed but molten salts and even helium gas might be able to meet the requirements. The unforeseen advantage that has now emerged is the simple fact that at 1 MeV, fission can take place in even isotopes, U^{238} or Th^{232} for example, and also in many of the actinides created by fission and capture, but previously considered to be nuclear waste. A well-designed fast reactor can operate at high temperatures and burn the even isotopes as well as waste. It is no surprise that many of the Generation IV proposals are based on fast reactor designs.

11.5 The Move to Modular Reactors, SMRs, and AMRs

A new concept emerged: factory-built and transportable modules, to replace the historical on-site construction. Small was now competing with large, with arguments that factory production of numerous smaller units would be much more economical than the large PWRs that are currently under construction. Sceptics still say that the economy of large scale will result in cheaper electricity and the small is beautiful initiative is, as yet, unproven in a commercial environment.

The answer to the question, 'How small is a small modular reactor?' is simply: small enough to get into a submarine. That is not quite the case, but historically, submarine reactors did require special designs because of space restrictions, and SMRs can take advantage of this experience. The modern versions of SMR are small in the sense that they must pass under bridges if they are transported by road, and the definition was originally intended to be not more than 300 MW of power. There is no precise definition of an SMR and small does not appear in the specification, they are linked to:

Modularity design.
Factory assembled and tested.
Road transportable.
Produced in large numbers for commercial advantage.

These four criteria are all related, and the design challenges are no longer simply the problems of nuclear physics, they are civil engineering design challenges with significant new problems associated with the on-site assembly of large modules. The term **AMR** is becoming more common, meaning **advanced modular reactor**, going beyond water as a coolant and operating at higher temperatures but otherwise an SMR.

There are many versions of modular reactors under development throughout the world, including the standard PWR type, high-temperature gas-cooled reactors, and molten salt reactors (MSRs) Two notable examples in the UK are the Rolls Royce SMR at 470 MW (e): a PWR using conventional UO_2 fuel at 4.95% enriched. The more recent development of a modular helium-cooled HTR by Cavendish Nuclear, going under the registered name 'U-battery', is concentrating on getting the modular design right and demonstrating that it can be factory constructed and assembled on site.

We have covered the generic changes that are suggested for Generation IV, so we can now examine some of the specific proposals in more detail.

11.6 Plutonium Breeding

Breeding was not taken seriously in the twentieth century it was experimental, with sodium coolant and fast neutron designs that required significant evaluation. The future availability of fossil fuels, with discoveries in the North Sea and elsewhere, suggested the extra effort required to breed was not yet worthwhile. Global warming has changed the emphasis considerably and breeding is now back on the experimental agenda. The worries about proliferation may well disappear with the introduction of new technology for reprocessing and it is suggested that breeding inside the reactor, not in a blanket, will produce plutonium with more Pu^{240}, making it unsuitable for use in weapons. The plutonium and thorium breeding cycles are both expected to develop into fully commercial concepts.

There is often some misunderstanding about the term 'breeding'. Plutonium breeding is as old as the first reactors, built at Hanford in the USA to produce plutonium for nuclear weapons. This was harvested for use in weapons. The many PWRs in operation also produce plutonium, but this is not harvested, it is burned in-situ, which means it is not usually referred to as breeding. Breeding with harvesting was pursued in the twentieth century, with several prototypes built and operational. Plutonium breeding comes from neutron capture in the plentiful U^{238}, it soaks up a large fraction of the 2.4 neutrons generated in the thermal fission of U^{235}. The U^{239} formed by neutron capture in U^{238} decays with a half-life of just 23.5 m to Np^{239}, which then decays with a half-life of 2.357 d to Pu^{239}. The three-stage process is:

$$U^{238} + n = U^{239} \qquad U^{239} (23.5\text{ m}) \text{ decays to } Np^{239} \qquad Np^{239} (2.36\text{d}) \text{ decays to } Pu^{239}$$

Some of the intermediate product Np^{239} will be removed by neutron capture, 32 b at thermal energy, since it remains in the reactor for several days. Neutron capture reactions in Pu^{239} continue to play a part in forming plutonium isotopes, Pu^{240}, Pu^{241}, and Pu^{242}, all being produced by successive neutron capture.

The optimal design of a reactor to produce a breeding ratio in the region of 1.0 requires fast neutrons. The UK developed a sodium-cooled fast reactor as a Prototype Fast Reactor, PFR, at Dounreay in Scotland. It produced power for the grid and had a plutonium fuel reprocessing plant on site. A Russian sodium-cooled fast reactor was successfully operated in Kazakhstan, between 1973 and 1993, and was said to have a breeding ratio of 1.2.

Even if a breeding ratio above 1.0 is not achieved, Pu^{239} is a fissile fuel in its own right. You simply can't avoid it in a uranium-fuelled fission reactor. It plays an important role in energy production as the burn-up of U^{235} progresses. Fission in plutonium can account for some 30 to 40% of the reactor power as PWR fuel comes to the end of its life. If plutonium is harvested and recovered by reprocessing it can be combined with uranium to form a mixed oxide (MOX). This has the dual advantage that Pu gives you more fissile material, and you are burning it to prevent it from escaping into the nuclear weapons camp. Plutonium breeding can be expected to be developed into a fully commercial operation in the coming decades, maybe with breeding in the core rather than the blanket, to produce plutonium with a weapon's safe isotopic composition.

11.7 Thorium Breeding

The thorium cycle has not yet been fully explored but it looks possible and has some attractive features. It has a very similar pattern to the uranium/plutonium cycle. Th^{232} is a monoisotopic fertile material and captures a neutron to form Th^{233}, which has a half-life of 21.8 m, leading to the intermediary protactinium, Pa^{233}, which has a half-life of 27 days, leading to U^{233}, a long-lived fissile material.

$$Th^{232} + n = Th^{233} \qquad Th^{233} \; (21.8 \; m) \; decays \; to \; Pa^{233} \qquad Pa^{233}(27 \; d) \; decays \; to \; U^{233}$$

Designing a reactor to achieve a breeding ratio greater than 1.0 is always difficult, and requires the best practices in neutron economy, but a fast neutron flux is not essential for Th^{232}, breeding can be achieved with a thermal neutron flux.

A very significant difference to the plutonium breeding cycle is the intermediary Pa^{233}, which exists in the reactor with a half-life of 27 d and has a large neutron capture cross-section. The thermal capture cross-section (not fission) is 40 b and the resonance integral is 850 b. If it is not removed from the reactor, much of it will have been destroyed by neutron capture before it has time to decay to the fissile U^{233}. One answer to this dilemma is to develop an online reprocessing system to remove the protactinium before it has any opportunity to capture a neutron. This is quite a challenge, and has moved the thorium breeding cycle, with harvesting, to a low priority in the eyes of many nations. New reprocessing technologies have been proposed and are under active development which, if successful, could lead to a renewed interest. However, it seems to make sense to develop the plutonium breeding cycle first, until the fertile U^{238} becomes in short supply.

Thorium is also suggested as an additive to other fuel without bothering about reprocessing and separation outside the reactor. It has been suggested as an additive to molten salt fuel with Pu^{239} as the fertile seed. Whatever U^{233} is produced in the reactor is burned in-situ.

A suggested benefit of the thorium cycle is that it may be more resistant to nuclear weapon proliferation because the U^{233} will be contaminated with U^{232}. This would be just as detrimental to weapon production as the presence of Pu240 is in the case of a plutonium weapon. U^{232} is a by-product from neutron irradiation of Th^{232}, produced after more than one nuclear reaction. Unfortunately, the intermediate product Pa^{233} has a half-life of 27 days giving plenty of time to remove it from the reactor, separate the Pa^{233}, and wait for it to decay without the presence of U^{232}. The critical mass for U^{233} is less than half of that for U^{235}, so it is not considered to be a safer bet and more resistant to nuclear proliferation.

Thorium as a fertile material is the subject of considerable interest in India and China since they both have large resources of thorium ore. They both have extensive programmes for the development of thorium in future reactors.

11.8 New Coolants

Gas coolants such as helium and carbon dioxide are self-explanatory. They are capable of operating at high temperature, they have no void coefficient to worry about or benefit from, and they have very little interaction with the structural materials they come into contact with. Helium will diffuse into metals to some extent, but it is an inert gas and has no chemical interaction. The only disadvantage of a gas coolant is its low mass density and inferior thermal conductivity. Removing a lot of heat quickly needs a large mass of coolant, which means that a low-mass coolant has to be pushed through the reactor at high speed. Nevertheless, Helium is very much on the agenda for the twenty-first century.

Liquid metals have been used before, with sodium and lead, or a lead/bismuth eutectic being the most common choice. These will continue to feature in some of the new designs.

11.9 Molten Salts

Molten salts are at the forefront of reactor development in the short term, with several **molten salt reactors, MSRs**, under advanced development. There are many options for the choice of salt composition, with the early version of lithium and beryllium fluoride **FLiBe** still considered to be an option. With so many salts available, it might be as well to avoid lithium since it breeds tritium, which is a radioactive hazard with a very high specific activity. The temperature range in an MSR can be extended right into the process heat region of 1,000°C, but the quest for higher temperatures, much higher than the 310°C that we are used to, may well present difficulties. Going up to 800°C takes all the structural materials into an environment where long-term survival, with unfamiliar redox potentials and radiation damage creating voids, particularly with fast reactors, will require many years of operational experience. Going up to 900–1,000°C may be a bridge too far, not that it cannot be done, but because it may reduce the operational lifetime of the engineering components so much that its shorter plant life makes it a poor investment. It will take 20 years to prove the long-term survival of structural materials at high temperatures. Very much a case of, the proof of the pudding is in the eating.

Novel features, such as placing the fissile material into the coolant, have been suggested. If a liquid form of uranium fluoride is introduced into the coolant, the liquid can be dumped into a containment well if the reactor gets into problems. However, pumping fissile material around in the coolant has the obvious disadvantage of contaminating the coolant pumps with fission product radioactivity and this may well override any other benefits. In some cases, the fuel can be changed online by real-time online reprocessing. However, one step at a time is the wise way to do it, and we still need much more experience with molten salts before we add the other complications, particularly if we aim to rise above 800°C.

An attractive option is to place the fuel in its own molten salt sealed-fuel-assembly, not that much different in concept to a conventional fuel assembly, except it is a liquid with internal convection and high temperature capabilities. It does not come into contact with a pump.

Moltex Energy Ltd specializes in the use of molten salts for their development of several new reactors. They started in the UK, expanded into Canada, and now have wide international interests. Their FLEX reactor is a novel concept—and FLEX is not an acronym, it is an abbreviation for 'flexible'. MoltexFlex, the UK-based arm of the company which is developing the FLEX reactor, has designed a reactor with sealed, but vented, fuel assemblies containing a mixture of sodium fluoride and low-enriched uranium (LEU) fluoride. The coolant is a different salt, a mixture of aluminium fluoride and sodium fluoride. The reactor is expected to operate at a temperature of 750°C for the primary coolant and a power of 40 MW (th). This is a surprising low power since wind turbines can now deliver 14 MW (e). There is another surprise, they may not generate electricity on the spot. They can pump the hot coolant into well-insulated massive heat storage tanks, containing a different choice of molten salt. The well-insulated tank remains hot it until the time is right for the production of electricity, with flexible timing. It is a GridReserve® thermal storage solution. This concept of flexibility is very attractive in the context of solar and wind electricity production since there will obviously be times when these alternatives to fossil fuel are not generating electricity. Even if they are generating electricity there is a commercial advantage in selling electricity at a time of your choosing, when prices rise during periods of peak demand. Even though the basic module is only 40 MW (th), a MoltexFLEX reactor plant can have any number of the mass-produced 40 MW (th) modules, all linked to several heat storage tanks.

Moltex is also developing, in Canada, an actinide waste burner (see Section 11.11). It is expected to be 300–500 MW (e), fuelled by a mixture of fissile material plus higher actinides recovered from spent nuclear fuel, and operating in a fast neutron spectrum. At the time of writing, this design had completed the first stage of the Canadian Nuclear Safety Commission Vendor Design Review; the design decisions are still in progress.

11.10 New Types of Fuel

The well-known UO_2 pellets are likely to be around for some time but new fuel concepts are under development and are in fact essential for some of the new designs. Ceramic pebbles and TRISO fuel, described in Section 6.8, are obvious options, with molten uranium salts likely to be one of the first deviations from UO_2. Changes of this nature, although ideal for reactor operations and safety, have implications at both ends of the fuel cycle, requiring new technology for fuel fabrication, reprocessing, and the logistics of spent fuel management.

11.11 Burning Waste and Using the Minor Actinides as Fuel

New initiatives are looking at schemes to burn the long-lived actinide waste with two incentives. It would reduce the amount of long-lived waste to be stored and the fissile worth of some of the actinides in the waste will make a useful contribution to the energy released for electricity production, killing two birds with one stone. It has obvious political attractions and studies are underway to confirm it would be commercially viable. It is ambitious and will require careful design of the nuclear reactor to successfully burn actinides in reasonable quantities. Unfamiliar long-lived actinides will be produced, and although it will be possible to burn more than you create, this initiative requires the development of new ideas for fuel fabrication and reprocessing. The design requires fast neutrons to take advantage of the fissionable worth of U^{238} and all the other actinides with both odd and even mass. Adding the value of the electricity to the savings of reducing disposal costs is considered by Moltex to be commercially attractive, a view clearly shared by the SNC-Lavalin, who announced a strategic partnership with Moltex in 2022.

11.12 New Reprocessing Technology

Reprocessing by means of the PUREX flow sheet has been the mainstay of the industry since its inception, so why do we need to change? Although there are many versions of the PUREX process the most common version yields a stream of very pure plutonium, which is considered to be a nuclear proliferation risk. The USA opposed reprocessing for this reason but are now supporting research into new flow sheets that are more proliferation resistant. This was announced in 2006 when the USA announced a **Global Nuclear Energy Partnership (GNEP)** to develop new proliferation-resistant recycling technologies.

Many of the Generation IV proposals call for a closed fuel cycle, with more selective options for the output streams. Breeding and burning of actinide waste are just two developments that will require reprocessing with proliferation-resistant features.

Reprocessing flow sheets use three different forms of technology:

Pyrometallurgy, which, as its name suggests, uses heat to initiate the separation of metals. This is a well-known method used for the smelting of copper and lead.
Electrometallurgy, a well-known method for the production of aluminium.
Hydrometallurgy, using aqueous solutions and chemistry to separate different elements.

The PUREX process is hydrometallurgy, but some of the new developments are electrometallurgical and sometimes referred to as pyrometallurgical since they also involve heat.

The name PUREX stood alone throughout the Generation III era but new names, UREX, without the P, denoting no pure plutonium stream, is the name of a new suite of flow sheets, UREX+1, UREX+1a, UREX+3, all with different **selective partitioning** goals. The objective is to support fast reactors in their endeavour to burn actinides and long-lived waste. It is also extremely desirable, some would say essential, to use selective partitioning of I^{129} and Tc^{99} from fission products destined for vitrification storage. Many other processes, DIAMEX-SANEX, COEX, and TALSPEAK are under development but are not expected to come into commercial operation before 2040.

The future of fission in the twenty-first century is ever-changing, with over-ambitious proposals being replaced by more realistic step-at-a-time approaches. A significant initiative in the development of reactors in the future is the Multinational Design Evaluation Programme (MDEP). It involves the International Atomic Energy Agency (IAEA) and 15 nations, with the objective of pooling information on designs, safety, and international design certification, a valuable contribution to speeding up the development of nuclear power. Several regulatory licencing organizations are also working together to specify and share international design certifications.

11.13 The Economics and Politics of Electricity Generation

This subject is the domain of accountants and politicians, which makes it a veritable minefield. The alternatives to fossil fuel all have a very large up-front construction cost and the conventional way for an accountant to deal with these costs is a straight-line amortization/depreciation of the assets over a number of years. It is common practice to reduce the assets at a linear rate over a period of not more than 17 years, which leads to a high cost in the early years and a low cost once the original capital has been written off the accounts. Fortunately, there is a meaningful way to deal with the large up-front costs when the lifetime of the plant is much longer. The **levelized cost of electricity (LCOE)** is given by calculating the total cost of everything during the lifetime of the project and dividing it by the total electricity expected to be produced when the capacity factor has been applied to the power capability of the plant. The total cost should include all the R&D costs prior to construction, as well as the mountain of safety assessments and general management charges that normally take place years before construction starts. The costs are usually broken down into capital costs, operating and maintenance costs, and financing costs. A very detailed LCOE is often provided by financial advisors to investment funds, and it would normally include a discount rate to deal with the return on capital investment.

Nuclear plants, built in the 1960s, were given a nominal lifetime of 25 years. Nominal because nobody knew how the structural materials would behave after prolonged bombardment by neutrons and gamma rays. The reality turned out to be about 40 years for the early reactors, with the new Generation IV SMRs predicting a lifetime of 80+ years. Consider the LCOE for a large two-reactor nuclear facility producing 3,200 MW (e) with a capital cost of $36 bn. If it is assumed the initial capital construction cost accounts for 66% of the total lifetime cost, then on the basis of a 30-year lifetime, the LCOE would be $0.07 /kWh. This figure has used a capacity factor of 0.90 and has ignored interest on capital. The retail price of electricity in the UK at the time of writing was about $0.40 /kWh.

An SMR producing 300 MW (e) can be built and operate for a total cost of $3.6 bn, which includes operational, waste management, and decommissioning costs. Assuming a lifetime of 80 years, it would yield a LCOE of $0.019 /kWh. This assumes a capacity factor of 0.9 but does not include a factor for interest on capital.

Solar farms occupy a lot of space, with 1 km^2 being required to produce about 40 MW of power. One of the largest solar farms in the world, the Bhadla Solar Park in India, occupies 56 km^2 and can deliver 2,245 MW. It stands in the Thar desert of Rajasthan, so one assumes the vast area it occupies has little commercial value and costs nothing. This would not be the case for solar farms in other parts of the world where land would have to be purchased or leased. The lifetime of solar panels is said to be about 30 years, but many panels built in the 1980s are still operating and the technology is improving all the time. The construction cost of solar farms accounts for about 88% of the total lifetime cost, with maintenance of the panels and the equipment to link into the grid accounting for the rest. The LCOE for the Bhadla farm, based on a lifetime of 30 years, a capital cost alone of $2.2 bn, and a capacity factor of 0.25 is $0.015 /kWh. Large uninhabited areas, all over the world, are available for building solar farms and long cables, or the production of hydrogen, make it transportable to urban areas where there is an immediate high demand for electricity.

A pioneering example of transporting electricity over a large distance is the massive Xlinks Morocco-UK Power Project. It has the potential to deliver 10,500 MW of renewable electricity over three undersea cables from Morocco to the UK. The electricity comes from solar and wind in Morocco. The total cost has been stated as $96 bn which, using an average capacity factor across both solar and wind of 0.3 would yield a 30-year plant life LCOE at $0.12 /kWh. This project is waiting for a political go-ahead since there are obvious concerns about the vulnerability of undersea cables and the long-term political relationships between the UK and Morocco. Nevertheless, it shows the potential for alternative energy supplies in the vast uninhabited regions of the world.

The LCOE for an offshore wind farm is usually based on a shorter lifetime of 25 years. The vertical structure and the blades are only fixed at one end, so they will vibrate in a high wind and eventually suffer from fatigue. The blades used in the high-power towers can be longer than a football pitch, so it is not surprising that the blades are locked in a stationary position in very high winds. There is not yet data on the fatigue lifetime of the large blades, but they can be replaced. An LCOE estimate for a wind farm designed to create 1,000 MW at a total cost of $2.9 bn, assuming a plant life of 27 years and a capacity factor of 0.5, yields LOCE = $0.03 /kWh.

It is difficult to evaluate the LCOE for fossil fuel since the LCOE is, by definition, intended to cover many years, usually decades. But because fossil fuel prices have fluc-tuated so wildly in recent years, the accepted value quoted is between $0.05/kWh and $0.18 /kWh, surprisingly larger than that claimed by solar and wind alternatives.

Accountants can manipulate the LCOE by using the options available to them to alter the parameters of the LCOE calculation. They can use concepts like opportunity cost and marginal cost as well as ignoring previous R&D and choosing interest rates and amortization factors to suit their own objectives.

Politicians have many ways to promote renewables by giving preferential VAT, tax rates, or tax credits. They can introduce legislation to benefit renewables and often give grants to support new technologies. Legislation to prohibit the use of fossil gas boilers in new-house builds is on the agenda for introduction in the next few years. In the USA, a switch to solar may give a 25% tax credit and their cleantech initiative gives tax breaks and subsidies worth a massive $369 billion. Governments can also use intangible arguments of national interest and assurance of supply to justify their decisions. You cannot put a price on assurance of supply; it means 24/7 coverage in all weather conditions and in the face of international turmoil. A significant initiative to support long-term investment in renewables is to offer **contracts of difference**. These are available in the UK but not legislated for in the USA. The contracts are sold by auction, using sealed bids, and can be considered as giving a guaranteed price for a limited period, maybe only seven years, but it helps large-scale projects to receive long-term investments. Governments may also promote a market for carbon, with carbon credits and carbon offsets being used to put fossil fuels at a commercial disadvantage to the alternatives.

11.14 The Utilization of $E = mc^2$

It has been a long journey since fission was discovered in 1938. It had an unfortunate beginning with an urgent need to produce Little Boy and Fat Man, but soon developed into the much more welcome realm of peaceful electricity production. The evolution faltered with Chernobyl and other disasters but seems to be back on track with global warming providing a global need for the future development of nuclear power. Mother Nature gave us $E = mc^2$ with its potential for the development of bombs, fission, and fusion reactors, and it is up to us to ensure it is only used for the benefit of mankind.

12

Nuclear Fusion

12.1 The Fusion Process

The sun is so hot that the thermal energy of the nuclei, buzzing around inside, is sufficient for them to come together and stay together, creating **fusion**. They have enough energy to overcome the electrical repulsion of the positively charged nuclei and are temporarily held together by the strong nuclear binding force before separating into the products of fusion. The sun operates at a mere 15 million °C, quite low for the fusion process, but the pressure in the inner regions of the sun is enormous. The pressure at the surface of the sun is less than planet Earth's atmospheric pressure but it rises to over 100 billion atmospheres at the centre of the sun. The 15 million °C temperature, working together with the high pressure, enables fusion in the sun to take place, albeit at a very slow rate. Fortunately, the sun is not in a hurry. Many fusion reactions are possible. The fusion of two deuterons D + D releases 3.268 MeV of energy but the outright winner for a fusion reactor here on earth, with more energy release and a much higher reaction cross-section, is when a deuteron (D) and a triton (T) come together and form a compound nucleus that decays into an alpha particle and a neutron:

$$D + T = He^4 + n, Q \text{ value} = +17.588 \text{ MeV}$$

The (D,T) reaction takes place with an energy release of 17.588 MeV; 14.1 MeV goes into the neutron, and the balance, 3.5 MeV, into the recoiling alpha particle. This energy release is much less than the 200 MeV released in fission, but the fuel mass for fusion is only 2 + 3 = 5, and the uranium mass for fission is 235, so the actual energy per kg of fuel favours fusion by a factor of 4.1. The number of neutrons produced in fusion, per MeV of energy released, is almost five times the number produced in fission since 2.4 neutrons are emitted per 200 MeV energy released in fission, compared to one neutron per 17.588 MeV of energy in fusion. The larger number of neutrons in fusion, together with the higher neutron energy, 14.1 MeV compared to 2.2 MeV in fission, means that the engineering materials in a fusion reactor are going to be much more susceptible to neutron damage.

The cross-section for the (D, T) reaction depends on the relative velocity of the particles. The thermal temperature to achieve a reasonable reaction rate needs to be above 100 million °C. The target for current experimental fusion reactors is about 150 million °C. Reaction cross-sections for several fusion reactions are given in Figure 12.1.

Uranium fission creates a lot of highly active fission products, some of them with an α decay half-life of thousands of years. The activity from fusion is mainly limited to the activation caused by neutrons being captured in the structural materials of the fusion

Understanding Nuclear Reactors. Brian Hooton, Oxford University Press. © Brian Hooton (2024).
DOI: 10.1093/oso/9780198902652.003.0012

Fusion Reaction	Cross Section	<av> cm³ sec⁻¹	at 150 Million ° K
D –T		1.9×10^{-16}	
D –D		2.0×10^{-18}	
D –He³		7.5×10^{-19}	

Figure 12.1 Fusion cross-sections. Since a temperature covers the Maxwellian spread of velocities, the cross section is given as an average over the velocity spread

reactor. A review of the periodic table shows that most of these activities have half-lives shorter than a year and it is reasonable to say that fusion will create much less long-lived radioactive waste than fission. Structural materials with very little activation have been developed specifically to minimize the activation produced by fusion reactors.

Fission is the source of energy in an atomic bomb, and fusion is the main source in a hydrogen bomb. It was not difficult to harness fission energy to produce electricity since the first fission reactor was connected to the national grid in the UK in 1958, only 13 years after the atomic bomb, but we know the quest for fusion power will take much longer.

12.2 Producing Fusion in the Laboratory

The first step in the journey is to establish, maintain, and control a plasma in the laboratory with conditions similar or superior to those present in the sun. The nucleons in the sun exist in the midst of a sea of electrons, not as a solid, liquid, or gas, but in a state of matter known as **plasma.**

In a fusion reactor here on Earth, high pressures cannot be produced, so it is necessary to go to temperatures over 100 million °C to achieve an acceptable rate of fusion. This requires producing a plasma in an environment where it doesn't come into contact with the walls of the reactor. The twentieth-century idea for the best way forward was to create it in a chamber the shape of a doughnut and prevent it from hitting the walls by carefully designed magnetic fields. The Russian tokamak soon became the favourite way forward when early measurements confirmed the tokamak design could achieve a temperature of 100 million °C. The many decades of development of the tokamak design since 1984, at JET (Joint European Torus), and elsewhere in the world, gradually dealt with the instability of the plasma, and we now confidently expect that we can ignite and contain a high-temperature plasma for many seconds.

In order to maintain a plasma, the first step is to ignite the deuterium and tritium gases, creating a plasma at a temperature capable of sustaining the burning. Once ignition is achieved the alpha particles, He⁴, produced by the fusion reaction will have enough energy to maintain the temperature. The ignition is achieved by three methods. Conventional electric heating by means of an electric current at low temperatures, microwave heating, and by the injection of energy in the form of high-velocity neutral particles. All three methods are used.

Once the plasma is ignited, the fusion products, neutrons, and alpha particles will interact with the surrounding engineering materials which, if they entered the plasma, may kill the fusion process by cooling down the plasma. Heavy nuclei such as iron and tungsten would not be in a plasma state because they would not be completely ionized, their inner electrons are so tightly bound. They would have a high degree of ionization, but the plasma would lose energy as it completed the ionization of the inner electrons. A light element, such as beryllium, with only four electrons, would be completely ionized and immediately become part of the plasma. The obvious place to worry about contamination of the plasma is the first wall facing the plasma. This is best made from a low atomic number element and the first candidate is beryllium. It has a high melting point, 1,287°C, and when it enters the plasma, it will itself become completely ionized and become part of the plasma. Unfortunately, neutrons interact with beryllium to produce helium which soon causes the beryllium to lose its strength. Choosing the material for the first wall is a compromise between beryllium, with a short operational lifetime, and the other option tungsten, which may well contaminate the plasma. The first wall of ITER (see Section 12.3) is expected to be coupons of beryllium up to 10 mm thick with a backing connection to a copper-alloy-heat-sink, cooled by water. This is an experimental facility, so a beryllium wall should survive the lifetime of the ITER programme. Other fusion reactors may well go for a tungsten first wall, in the belief that tungsten contamination of the plasma will be manageable. If the first wall is made from tungsten with a melting point of 3,422°C, it will also benefit the breeding of tritium since there will be neutron multiplication by the (n,2n) reaction in the tungsten isotopes. The first wall is the ideal spot for intercepting 14 MeV neutrons; neutron multiplication anywhere is good news for breeding tritium.

Inside the plasma only a small amount of the fuel becomes fused to generate power; most of it needs to be extracted and recycled several times. The fusion reactor needs a mechanism for removing spent plasma for recycling. This includes the removal of the helium produced in the fusion reaction. This helium is sometimes referred to as the 'ash'. The engineering unit to achieve this is called the **divertor** and since it will be subjected to a direct hit by the extracted plasma it will get very hot. Fortunately, it is some distance away from the plasma, so contamination from the divertor into the plasma is manageable. The higher temperature of the divertor means it will be constructed in tungsten.

Any tokamak, operating to produce electricity and to breed its own fuel, will require superconducting magnets configured to prevent the plasma from hitting the walls. Also surrounding the toroid will be the blanket containing all the engineering pipework for getting the heat out to generate electricity and lithium to breed and extract the tritium fuel, essential for a feasible continuous operation. There are many engineering materials involved and research has been carried out to find new alloys that will meet the stringent requirements of the environment in the fusion reactor, yet still have very low activation into radioactive products that will need to be managed during decommissioning. EUROFER97 is one new alloy designed specifically for use in a fusion reactor. It has been tested for mechanical and microstructural properties in ten variants of the elements, Cr, Mn, V, W, Ta, and Si, but tests on other materials, with even lower activation, are ongoing.

The lifetime of the structural components in a fusion reactor is not expected to be long and provision for replacement during maintenance periods is planned to take place using robotic assistants to complete the tasks. The maintenance will need robotics since the work will be carried out in a high-radiation environment. The precise frequency and need for removal of components will only become known in the light of operational experience.

Producing a plasma in the laboratory with similar characteristics as the sun has proved to be a very difficult challenge but the light at the end of the tunnel appears to be in sight. Although there have been many decades of disappointment and slow progress, the understanding of plasmas has grown to the extent that we now believe we have a good understanding of how to control and maintain plasmas. Several tokamaks, the Korean KSTAR (Korea Superconducting Tokamak Advanced Research) and the Chinese EAST (Experimental Advanced Superconducting Tokamak) have been working in parallel with JET and providing valuable experience in the run-up to ITER. Both of these tokomaks now claim to have exceeded 100 million °C for a duration exceeding 20 seconds. The next significant major milestone in the journey will be the commissioning and operation of the **International Thermonuclear Experimental Reactor (ITER).**

12.3 ITER

The building of a truly international collaboration, ITER, is underway in southern France. Some 35 nations are involved in the building of the world's largest fusion experiment, a tokamak. It is intended to maintain a plasma for a long time and produce a positive net energy, more energy out than in, but there is a lot more engineering involved to recover the energy of fusion and turn it into electricity for the grid. ITER is an experimental facility that will not generate electricity, but it will test the removal of the heat using helium or steam as a coolant and investigate several options for the design of the **breeding blanket**, essential for commercial operation.

Deuterium/Tritium plasma operations are not scheduled to start until 2035 and it will probably be after 2050 before we see DEMO, a full-scale demonstration of an electricity-producing fusion reactor. The DEMO phase is not intended to be an international collaboration and nations are already designing their own versions.

12.4 MAST and STEP

One of the first variations away from the doughnut-shaped tokamak was the MAST (Mega Ampere Spherical Tokamak) development at Culham Centre for Fusion Energy

(CCFE), in the UK. It was based on a shape like a cored apple, almost spherical, and often referred to as a spherical tokamak. The new shape is more compact and uses different magnetic confinement designs. MAST worked so well that an upgrade was initiated and is due to operate in the 2020s. The upgrade will include the testing of a new type of divertor, the Super-X-Divertor, which is expected to reduce the heat taken up by the divertor and enhance its lifetime in the reactor by a factor of ten. This is an important area for improving the commercial viability of fusion reactors.

The MAST upgrade will be a proving ground for the basis for STEP (Spherical Toka-mak for Energy Production), a comprehensive programme to create a working fusion reactor that is designed to go all the way, sending electricity into the national grid by 2040. It is due to be built on a chosen site at West Burton, Nottinghamshire, UK. It is funded by the UK Atomic Energy Authority.

STEP is being designed to produce between 1,000 MW to 2,000 MW (th) but may only generate some 100 MW of net electrical power for transmission into the grid. These figures are a reminder that the economics of fusion reactors are still uncertain since all the factors of economic significance will not be known for decades. Fusion reactors will require a large overhead of electrical power to drive all the ancillary plants. These include vacuum pumps and refrigeration for all the aspects of the fusion fuel cycle. Superconducting magnets don't require as much energy to drive them, as the name suggests, but they operate at very low temperatures, and they require cool-ing. Additional energy in the form of electrical heating or by means of neutral particle injection is also required to ignite the plasma, a fire lighter. There are significant fuel costs in managing both the tritium and deuterium inputs, but the big uncertainty is the cost of maintenance when the first wall and other components need replacing at the end of their operational life. These overheads all indicate that a truly commercial plant will have to be a large one.

12.5 The Fuel for Fusion

The fuel for fusion power is deuterium and tritium. Deuterium is present in water as a 0.015% component of hydrogen. It will require isotopic separation, but it is much simpler to separate a mass of 1 from a mass of 2 than it is to separate 235 from 238 in the case of uranium. Tritium is not a natural isotope in nature, it is radioactive with a half-life of 12.3 years, and needs to be 'manufactured'. The plan for fusion reactors is to establish an opening inventory of tritium, for the very first fusion reactor, but fresh fuel for the ongoing operation of the first reactor, and for the opening inventory of any additional reactors must come from breeding fresh tritium in the blanket.

The fuel for the first reactor's opening inventory will be obtained from tritium produced in fission reactors. This is a well-established process since it was used to supply the material for the hydrogen bomb. Some reactors use heavy water, D_2O, as a moderator and cannot avoid producing some tritium from neutron capture in deu-terium. This is extracted on a regular basis and will provide a large contribution to the opening inventory. A fusion reactor with a thermal power of 1,000 MW will require 154 g of tritium per day. This is based on the assumption that fusion releases 17.6 MeV of energy, but it ignores the fact that some energy will be lost as neutron multiplication takes place, and some will be gained when a neutron is absorbed in Li^6 to breed new tritium. Nevertheless, 154 g/day is quite a large amount of tritium fuel, so let us see how much we might be able to produce from a fission reactor if we needed to do so! A fission reactor operating at a power of 1,000 MW (th) will produce plenty of neutrons. The neutron balance in the reactor suggests that, whilst still maintaining a chain reaction, perhaps 10% can be made available for breeding tritium. This could lead to the production of 3.24 g of tritium per day, which is about 2% of that needed for a fusion reactor of similar power. The conclusion is that breeding is a must, fusion reactors must breed their own fuel and a breeding ratio considerably greater than 1.0 will be required for the commercial viability of fusion reactors.

12.6 The Tritium Breeding Ratio (TBR)

The tritium breeding ratio (TBR) is defined as the number of tritons produced per sec in the blanket divided by the number consumed per sec in the plasma. It needs to be greater than 1.0 for two reasons. It provides the fuel to keep the fusion reactor itself going, but it also needs to create a surplus to provide the opening inventory for any new reactors. Tritium management is a considerable challenge, not only because it is a notable radioactive hazard but because every gram needs to be recovered in a timely manner. There will be some 'losses', not to the environment, which is absolutely forbidden, but probably in the form of 'hold up' in the tritium processing plant, with some absorption in metal surfaces and solid compounds hiding in crevasses. There is also the natural loss due to tritium's 12.3 y radioactive decay half-life, which is unavoidable.

Each fusion reaction that consumes a triton generates just 1 neutron, and only one, so how is it possible to get a breeding ratio greater than 1.0? The answer lies in the neutron multiplication process of the (n,2n) reaction. Extra neutrons can be created by means of the Be^9 (n,2n) Be^8 reaction or a similar process in the lead isotopes. Other nuclei are capable of the (n,2n) reaction but Be and Pb stand out as the best neutron multipliers. One neutron enters beryllium, which is monoisotopic, Be^9, creating 2 neutrons and Be^8, which immediately disintegrates into two alpha particles.

$$Be^9 + n = n + n + He^4 + He^4$$

The Q value for this reaction is –1.573 MeV, meaning the neutron must have at least 1.573 MeV for the reaction to be possible. It is this reaction in a first wall made from beryllium that would generate helium and soon make the wall fail in its structural capability. The other main candidate for neutron multiplication is lead, Pb, which has several isotopes, all capable of enabling neutron multiplication by the (n,2n) reaction. The Q values are all similar, with a weighted average of –7.5 MeV. The probability of these reactions is higher in Pb than Be, but because Pb is much heavier than Be there are fewer Pb nuclei per unit volume. This means the mean free path, the mean distance travelled to produce a reaction, is much the same for both materials. Be and Pb are both candidates as neutron multipliers for experimental investigation during ITER operations.

When the neutrons enter the blanket, they will undergo interactions similar to those in a fission reactor, scattering, moderation, and various interactions with all the nuclei in the blanket. The neutrons will breed tritium by interacting with lithium in the blanket. Li^6 captures a neutron and creates tritium by means of the reaction:

$$Li^6 + n = He^4 + T, Q = +4.78 \text{ MeV}$$

This positive Q value means the reaction can take place at all energies and the thermal neutron cross-section, 936 b, is particularly large, which means that most of the neutrons that don't escape will in fact create tritium.

The other Li isotope Li^7 also plays a part in tritium production by means of the reaction:

$$Li^7 (n,n') = He^4 + T, Q = -2.468 \text{ MeV}$$

The reaction $Li^7(n,n')$, inelastic scattering, leading to an excited state in Li^7, is followed by the emission of an alpha and a triton. The inelastically scattered neutron is still alive, and since we are starting with 14 MeV neutrons and the Q value for this reaction channel is −2.468 MeV, we could get more than one inelastic scattering out of a neutron before it gets below the threshold. This would, in theory, give a possibility of three or even four tritons from just one neutron travelling into the blanket. When the neutron energy is below the threshold it will still be available for creating an additional triton if it survives and ends up being absorbed by Li^6 at thermal energy. This Li^7 process can be viewed as a multiplication effect since one neutron can create several tritons. It can be viewed as an alternative to relying on neutron multiplication in Be or Pb and maybe strong enough, on its own, to yield a breeding ratio significantly greater than 1.0.

Neutron multiplication in Be, Pb, and W are all available as options to provide additional neutrons for tritium breeding. Li^7 is also capable of producing tritium directly, so it will be the presence of Be, Pb, W, and lithium in the design of the reactor and its blanket that will determine the TBR.

Monte Carlo calculations, using the nuclear data for (n,2n) reactions and the breeding reactions in lithium, suggest that a breeding ratio in the range 1.1 to 1.2 is achievable. We can remain optimistic, with adequate justification, that an acceptable breeding ratio can be achieved, but we will have to wait for ITER to know the truth.

Producing tritium in the blanket, even with a large TBR, is not the end of the story, far from it. Tritium processing and recovery is essential to provide adequate amounts of it as fuel. Tritium, as a hydrogen isotope, will diffuse into anything it comes into contact with, particularly at high temperatures, and it is very reactive, combining with many elements to form hydrides and other compounds. It will mix with the other hydrogen isotopes losing its purity as tritium, so a fusion reactor needs a very efficient tritium processing plant as part of the immediate operations to keep the plant running. Solids containing tritium can get temporarily 'lost' or 'hidden' in the processing plants that are necessary to produce tritium for reintroduction into the plasma. Tritium needs to be recovered and processed as it is bred in the blanket and rescued from the reactor, as it contaminates the walls and the divertor during the recycling of the exhaust gases and ash. It must be true that tritium cannot be lost to the environment since it is a very serious radioactive hazard with a high specific activity due to its short half-life of 12.3 years. Nevertheless, some of it is bound to be 'held up' in the processing plant and this may well lead to a logistical problem of providing the right amount of fuel in a timely manner. Tritium processing is a significant challenge over and above the breeding ratio.

12.7 Venture Capital

The tokomak, with its magnetic confinement route to fusion, dominated the twentieth century and is still the front-runner today. ITER is benefitting from decades of research and experience at JET and many other similar facilities all over the world. There is considerable, and justified, optimism that ITER will demonstrate a well-controlled plasma at a high enough temperature, pressure, and duration to make a considerable net energy gain. This would lead to the next stage of producing electricity and feeding it into the local grid. At the moment we have no idea of the commercial cost of producing electricity by fusion. This route continues to be supported by the whole of the international community but there is no shortage of sceptics, and private enterprise as well as national laboratories have been busy looking at alternative routes to achieve fusion.

Some of the alternatives have adopted variations on magnetic confinement and others are trying inertial confinement to achieve fusion. I think it is fair to describe them all as speculative, but who knows which approach, or combination of approaches, will achieve the best result. Not all of them are committed to the D + T reaction, with D + He^3 and the B^{11} + p reaction making a surprise appearance on the stage. Venture capital is a very appropriate expression! A list of some of the organizations that are benefitting from venture capital is given in Figure 12.2.

12.8 The Conclusion on Fusion

There is a good chance that the technical objectives of ITER will be met, and the reactor will produce more energy than it uses. Technical success does not always equate to commercial viability and commercial success is still uncertain due to the many incidental cost factors associated with the fusion reactor fuel cycle. Natural lithium is about 7.5% Li^6 and 92.5% Li^7. Enrichment in Li^6 will probably be required to achieve a high enough TBR but the enrichment process is well established, the COLEX method, named from 'column exchange'. It involves the chemistry of mercury amalgams and

ORGANIZATION	TECHNOLOGY
STEP (Spherical Tokamak for Energy Production) UK .Gov.	2040s Target, Smaller and more compact field design
Tokamak Energy, Oxford UK	Shaped like a cored apple, new magnetic configuration, staged development
CT Fusion Seattle, Spheromak, Dynomak™	A new and cheaper way of magnetic confinement
Fusion Reactors Ltd. UK	Magnetic confinement design evaluation
Princeton Fusion Systems. USA	A microreactot using $D + He^3$ fuel. A portable sealed unit
Lockheed Martin USA	A small portable compact design
Large Helical Device, Japan	A stellarator (twisted doughnut) design
LPPFusion, New Jersey USA	Brand new ideas, Focus fusion using hydrogen and boron fuel and the $B^{11} + P$ reaction
Sandia National Laboratories, USA	Development of the Z machine concept, Inertial crushing of fuel
National lgnition Facility, California USA	Laser focused compression of fuel
HB11 Energy, Australia	Laser fusion of hydrogen and B^{11}
Marvel, Munich Germany	Laser fusion of hydrogen and B^{11}

Figure 12.2 Venture capital alternatives to conventional Tokamak fusion

lithium hydroxide, with Li^6 becoming preferentially concentrated in the amalgam. The process has been used to take the Li^6 enrichment to well over 50%, but it does add another cost to the fuel cycle.

We spent 50 years in the twentieth century learning how to produce a sustained plasma, the first 50 years of the twenty-first century may lead to understanding the engineering problems and getting more energy out than we put in, but the difficult part will be producing electricity at a commercially viable cost and who knows how long that will take!

We are trying to produce fusion energy here on earth, by simulating the conditions on the sun, but we can at least benefit from the fusion that takes place in the sun itself, at a distance of 98 million miles away. The sun's energy arrives in uninhabited regions, the deserts, the bush in Australia, and the vast areas of Siberia at an amount of 1.366 kW m^{-2}, the **solar constant**. These areas are enormous and are all capable of harnessing solar energy to sell at a very competitive price. They do not have a radioactive waste problem or significant decommissioning costs, with the added advantage that there is no one to object and say—not in my backyard. The much simpler technologies of solar and wind may yet win the day completely.

13

The Hydrogen Strategy

13.1 The Basic Properties of Hydrogen

It would be somewhat remiss to talk about energy in the twenty-first century without mentioning hydrogen. The hydrogen economy has been a topic of conversation for decades and there is now the possibility that it will become a revolution, playing a significant role in energy supply. Historically, hydrogen has been making a large contribution to the world's economy for more than 200 years; the current global use exceeds 70 million metric tonnes per annum, almost all of which uses hydrogen as a chemical, not an energy source.

This is the story of hydrogen, what it is, how to make it, what it costs, and what it is used for. It promises to be very helpful in the battle against global warming since when it burns, it combines with oxygen to form water, with no creation of carbon dioxide. This new hydrogen strategy is expanding rapidly into energy supply, and if it takes off it could replace many of our traditional uses of fossil fuel. It has already made inroads into transport by means of fuel cells and there are those who see it as being delivered by existing pipelines into our homes. The main criterion for the revolution is going to be the cost to produce it.

Since energy, electrical or otherwise, is needed to produce hydrogen in the first place, its role in the fight against global warming is quite subtle. Hydrogen, produced using electricity from solar panels, may well be used to boil our kettles at home. This must be less efficient than the direct use of the same electricity to boil the kettle! However, there are several scenarios where the use of hydrogen as an intermediary makes sense, and this chapter will explain just how hydrogen will have a role to play in our fight against global warming. It looks at all the aspects of the hydrogen strategy and will allow you to form your own opinion on the future role of hydrogen in the world economy.

Hydrogen was discovered by Robert Boyle in 1671 but it was not named until Henry Cavendish isolated it as a discrete substance in 1766. It is the most abundant element in the universe, present in our everyday lives in the form of water, but not as a gas since it combines with other elements to form compounds. There are trace amounts in the atmosphere, at a level of less than one part per million, and very small amounts in underground rock formations. It has the lowest density of any gas, being only 0.000082 g cm^{-3} at ambient temperature and pressure.

When it combines with oxygen to form water it releases a large amount of energy, per gram, more than any other substance. This gives it a role as a direct substitute for

Understanding Nuclear Reactors. Brian Hooton, Oxford University Press. © Brian Hooton (2024).
DOI: 10.1093/oso/9780198902652.003.0013

Fuel	Gross Calorific Value Mega Joule per kg
Hydrogen	141.7
Ammonia	22.5
Methane	55.5
Natural Gas (US Market)	52.2

Figure 13.1 The calorific value of various fuels

fossil fuels. The gross calorific value is compared to some of the other common fuels in Figure 13.1.

It is not particularly dangerous when compared to other fuels. Petrol is much more dangerous since it is heavier than air and a spillage gives it the opportunity to mix with air to form a highly explosive mixture. Natural gas and town gas, manufactured from coal, have been used in our homes for ages and are not considered to be an unmanageable hazard. Hydrogen has the added benefit that if there is a leak it will rise like a rocket, due to its low density, and not remain concentrated enough to explode.

The fact that hydrogen is a very low-density gas means that it occupies a lot of space at normal pressure, making storage expensive. The solution to the storage and transport problems is to store at high pressure, up to 700 times atmospheric pressure, or convert it into a liquid. The high-pressure option is the most common, with gas cylinders being transported by trailer. Liquefaction is also a possibility, and does take place, but requires a very low temperature since the boiling point of hydrogen is −253°C. It can also be stored and shipped by converting it into a compound, such as ammonia, NH_3, or lithium hydride, both of them occupying much less space per gram of hydrogen, but the conversion process to make the compound, and the eventual cost to recover the hydrogen, inflate the price.

Hydrogen has been around for a long time. One of the main products is the production of ammonia, NH_3, for agriculture fertilizer. The oil industry uses large quantities to remove sulphur and also to produce aviation fuel and diesel by the hydrocracking process. The third main traditional user is the plastics industry which uses hydrogen to make cyclohexane and methanol (CH_3OH), both of which are intermediaries for the production of plastics. A more recent use is as a propellant for space rockets. Hydrogen is also used in many other commercial processes, often as a reducing agent to convert metal oxides into pure metal. Most of these markets are using hydrogen as a chemical and if we are to have a real revolution it will require hydrogen to be produced and used as a source of energy.

13.2 The Production of Hydrogen

The traditional way to produce hydrogen is by passing an electric current through water with the energy of electrolysis separating the hydrogen and oxygen content. The equipment to do this is called an electrolyser; an example is shown in Figure 13.2.

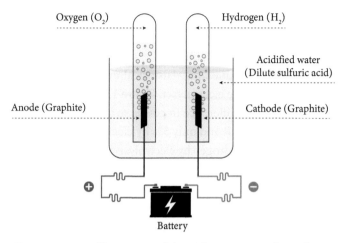

Figure 13.2 An illustration of electrolysis. It is used to split water into hydrogen and oxygen

The early versions used a liquid as an electrolyte, but modern technology now has the option of using a polymer electrolyte membrane (PEM) or a solid oxide electrolyte cell (SOEC). The PEM and the SOEC methods are capable of handling a higher current giving rise to a higher yield. Production by electrolysers was often linked to dedicated hydroelectric power plants operating 24/7.

Most of the historic hydrogen production has not come from an electrolyser, it was generated from hydrocarbon feedstock using steam heat processing in the region of 700–1,000°C. At the forefront is **steam methane reforming (SMR)** (not to be confused with small modular reactor, also SMR), which uses a nickel-based catalyst and high-temperature steam. Methane, CH_4, has four atoms of hydrogen per molecule, and they are released by the reaction with steam according to:

$$CH_4 + H_2O = CO + 3H_2$$

This stage leaves the carbon as CO, not considered to be a global warming gas since it has a comparatively short atmospheric lifetime. It still has a detrimental effect on the environment since CO reacts with hydroxyl (OH) radicles in the atmosphere which, in turn, affects the concentrations of ozone and several problematic greenhouse gases. Although SMR uses methane, there are many variations, capable of extracting hydrogen from coal, biomass, plastic waste, and other hydrocarbons.

The SMR can be followed by a second stage to produce additional hydrogen, the **water–gas shift reaction**. It's an exothermic reaction releasing hydrogen at the lower temperature of 360°C.

$$CO + H_2O = CO_2 + H_2$$

Since the water–gas shift reaction produces CO_2, it should be accompanied by **carbon capture and storage (CCS)** to prevent the CO_2 from getting into the atmosphere.

The extra cost of this additional stage should be included when comparing the cost of different methods of hydrogen production.

This concludes the technical account of the two common methods to produce hydrogen but leaves us with a dilemma. Historical evidence suggests that the MSR method costs much less than the electrolyser. Figure 13.3 shows the cost variation for several production paths. The difference in cost is so large that historical production has been split into 95%+ by SMR and only 5% by electrolysis.

Since the cost of producing hydrogen by electrolysis is so much higher than by SMR, it would seem to indicate, at first glance, that there would be no buyers for hydrogen produced by the electrolysis method, it's too expensive. Fortunately, there is a simple way to reduce the price to the customer. You use a different accountant. Selling at marginal cost, which is setting the price of a product to equal the extra cost of producing an extra unit, is a common way to do business. It often means using only the cost of materials and labour and avoiding overheads and other indirect costs. Another reason why electrolysis is attractive is because hydrogen is a form of energy storage, and when the sun is shining but there is no local demand for electricity, storing the energy by the production of hydrogen by electrolysis can be justified. There is an enormous potential for the production of green hydrogen using solar and wind energy when there is no demand for electricity on the spot. In this situation, the lower cost of producing hydrogen by SMR is irrelevant. The amount produced by SMR is likely to increase since the market for using hydrogen in fuel cells is growing rapidly, and in 2021 the US Government announced a mission to reduce the cost of hydrogen production by as much as 80% in a decade. They aim to bring the cost down to **$1** per **1 kg** in **1 decade.** They call it the '111' initiative.

The future may well see new methods to produce hydrogen using much higher temperatures than are used today. **Methane pyrolysis** is a direct process to separate the carbon and hydrogen in methane at a high temperature. It proceeds in a simple manner according to:

$$CH_4 = C + 2H_2$$

The Production Process	US $ Cost per kg of Hydrogen
Coal Gasification No Carbon Capture and Storage CCS	1.34
Steam Methane Reforming SMR No CCS	2.08
Steam Methane Reforming SMR With CCS	2.27
Electrolyser using Wind/Solar	6.0 +

Figure 13.3 Hydrogen production costs. From the International Energy Agency, 2019

The method requires a molten metal catalyst but does not generate the greenhouse gas CO_2. The carbon produced is in granular form and can be buried or used as manufacturing feedstock. This technology has some way to go before becoming a reality, but BASF, one of the major manufacturers of hydrogen, is well on the way to operating a pilot plant.

Water splitting by temperature alone can be used to generate hydrogen without any need for carbon capture, but, without a catalyst, it requires a very high temperature in the region of 2,000°C. If it is carried out with the benefit of a catalytic thermo-chemical cycle it will work in the temperature range 600–1,000°C. The hope is that the two high-temperature methods, pyrolysis and direct water splitting, can benefit from nuclear reactors, developed to operate at such high temperatures. They would produce hydrogen first, and then go on to also produce electricity. Generation IV reactors, operating at high temperature and 600 MW (th), are said to be capable of producing 2 million cubic metres of hydrogen per day.

13.3 Carbon Capture

Carbon capture by nature, sequestration, in trees, grasses, and oceans, has controlled the CO_2 cycle since the beginning of time but is not sufficient to keep CO_2 levels down to what is required to combat global warming. The release of CO_2 due to burning fossil fuel can be subjected to carbon capture technology to prevent it from being released into the atmosphere, and it looks as though the development of this technology is essential.

Carbon capture and utilization (CCU) captures carbon and converts it into a chemical form suitable for sale and commercial use. There is a market and a price for carbon compounds, but this option feeds a market which is already well supplied, so making CCU commercially viable may be easier said than done. The alternative, **carbon capture and storage (CCS)** is preferable, but it still involves a processing cost and also finding a long-term storage solution without problematic environmental factors.

CO_2 can be removed from the atmosphere itself and also at its point of origin, in flues where fissile combustion is taking place, or where it is created in chemical plants. The gas can then be injected into geological formations deep below the Earth's crust. There are many deep underground chambers, with enough capacity to accommodate the entire problem. The oil giants were the pioneers in this technology since injecting CO_2 into oil wells enhanced their output. The oil companies had a clear commercial case for carbon capture and injection since the extra yield from a well would more than cover the costs of the injection. The commercial incentive to producers of CO_2 to capture and store, with the added cost of doing it, is unclear. They are being asked to pay for a process they don't need. The alternative, to them, is simply to release CO_2 into the atmosphere, which costs nothing. It will need government intervention with incentives, or legislation, to persuade producers to adopt CCS. An alternative to gas storage is to produce water saturated in CO_2, sparkling water, and inject it into specific rock formations that will convert the CO_2 into a solid carbonate compound over a period of just a few years. Several countries are using this method of underground storage.

New CC technology is being introduced by venture capital startups. UNDO, short for undo CO_2, uses a powdered form of basalt rock, magic dust, scattered over a large area of sloping land to capture the CO_2 in rain when it falls. Rain absorbs CO_2 to form carbonic acid, acid rain, which combines with the basalt to form compounds of carbon. The process is completed by the carbon compounds being washed down the slope into the nearby ocean, where they sink to the bottom out of harm's way.

13.4 Energy Storage

Energy storage technology is an important topic in its own right. Water storage in reservoirs is an obvious method, with reservoirs being specially engineered to receive pumped water from power stations during night-time, when domestic demand is low. Batteries are probably the most familiar form of energy storage, with modern technology introducing lithium batteries as one of the best from the power-to-weight ratio point of view. An alternative to lithium is the nickel hydride battery with a potassium hydroxide electrolyte. It may be cheaper to produce for a static storage facility where space and energy storage per kilo is not as important. There are numerous novel technologies, such as storing energy in a heavy flywheel operating at high speed and several examples of lifting weights to great heights and recovering the energy when they fall under gravity. Compressing a gas or liquifying it and then allowing it to warm up and come back to ambient pressure is the basis for several operational energy storage systems. However, hydrogen itself has an intrinsic property as an obvious energy store, it stores sunshine, either as a large mass of combustible gas under pressure or as a liquid.

13.5 New Markets for Hydrogen

Most of the traditional markets for hydrogen used it as a chemical, not to produce electricity or to burn it for heat. New markets for hydrogen are emerging so fast that the demand may well exceed supply and inflate the price. We can start to appreciate the influence of hydrogen on our lives by considering a few examples and talking about them as though they are case studies.

The transport sector is already very familiar with hydrogen fuel cells. They work like electrolysers but in reverse. Electrolysers use electricity to split water into hydrogen and oxygen, fuel cells combine hydrogen and oxygen to form water with the production of electricity. They are becoming the first choice of energy for forklift trucks in the USA, with the number of cells in use growing by more than 10% a year. Globally, the market for fuel cells is expected to reach 5,600 MW by 2030.

A new market for hydrogen is its use as a combustion fuel in engines for heavy vehicles. JCB, an international leader in earth-moving equipment, bulldozers, has developed an alternative to its diesel engines. It uses pure hydrogen as a fuel with spark plug ignition, pistons, and the like. They have produced an excellent replacement for their diesel engine. JCB is not just optimistic about its adoption—it has done it. Hundreds of engines have been produced over several years, and the final specification is undergoing endurance trials before they become the norm in JCB earth-moving equipment. The hydrogen is stored under high pressure and the fuel gauge just reads the reduction in pressure as the fuel is used. The technology is complete, but we need to keep our fingers crossed and hope that there is an assured supply of hydrogen.

Every revolution requires leaders, and Australia is the first in the world to transport liquid hydrogen, at –253°C, by sea to an international market. The Japanese word for hydrogen is 'suiso', and the ship, Suiso Frontier, sailed from Australia to Japan with a small cargo of liquid hydrogen for the first time in January 2022. Australia makes it from their abundant supplies of brown coal, lignite, and hence the name 'Brown Hydrogen' is given to these shipments. The colour can be said to be effectively blue through the use of carbon offsets. The pilot stage produced only 3 tonnes of hydrogen, but the long-term aim is to increase production to over 20,000 tonnes with CCS depositing the CO_2 in an offshore reservoir.

Australia has become a significant investor in solar energy. It has plenty of sunshine to go with its large unpopulated interior, and no one likely to object. The Murchison Hydrogen Renewables Pty Ltd endeavour, located near Kalbarri in West Australia, will use 700 wind turbines (3.7 GW) and a solar PV farm (1.5 GW) to produce green hydrogen. It will be produced from demineralized water using electrolysers. It also has a large lithium battery storage facility to store the energy when it is surplus to immediate needs. The hydrogen will be used to manufacture ammonia at a rate of 2 mega tonnes per annum and move it by a cryogenic pipeline to an offshore marine export facility for shipment to Japan. This is a good example of the subtlety of the hydrogen strategy. Solar energy is being used to produce ammonia which was, hitherto, produced using fossil fuel, thus contributing to global warming objectives.

The NEOM Green Hydrogen Project is due to be commissioned in 2026. It is located in Saudi Arabia and designed to produce 3,900 MW of onshore wind and solar. It will also have its own energy storage facility. The power produced is roughly the equivalent of four large nuclear power stations, each 1,000 MW I, but don't forget the capacity factor, explained in Chapter 1, which takes into account the hours of darkness and the time when the wind is only a gentle breeze. A typical combined capacity factor for both solar and wind in Saudi Arabia is 25%. These two examples, both having energy storage facilities, emphasize the need to recognize that energy should not be allowed to go to waste when it cannot be used immediately.

The next example is the Worley design for a 36,000 MW electrolyser plant on an artificial island off the coast of the Netherlands. It has the power equivalent of 36 large nuclear power stations; how ambitious can you get? Worley is Australia-based but has alternative energy projects all over the world.

These projects are typical of the changes that will take place as the hydrogen strategy emerges. They are currently concentrating on using solar and wind energy since the Generation IV nuclear reactors have yet to arrive. Any nuclear reactor could be used to produce hydrogen, from the electricity it generates or from the benefit of the high-temperature process heat that some future reactors could deliver. There is virtually no electricity generation from hydrogen fuel at the moment, but hydrogen-fuelled gas turbines are on the horizon and if there is an adequate assurance of supply, and the price is right, it will happen.

13.6 Hydrogen in the Colours of the Rainbow

Hydrogen is a colourless gas, yet it finds itself described in all the colours of the rainbow. The only exception appears to be 'Indigo Hydrogen', which was a pop song by the guitarist Siyuma in 2019.

The colours are now linked to the method of production with green hydrogen, the most desirable colour, being made by electrolysis from clean electricity coming from renewable sources such as solar or wind. Yellow hydrogen is the colour used for production, specifically using solar cells.

The name blue hydrogen is associated with hydrogen produced from hydrocarbons, such as methane, by SMR, but this colour requires the CO_2 by-product to be dealt with by CCS, and not released into the atmosphere.

Grey hydrogen accounts for the bulk of hydrogen produced today. It comes from methane or natural gas, without any form of carbon capture. The darkest forms of hydrogen are black and brown, made from black or brown coal.

Not strictly a rainbow colour, pink hydrogen is the name given to electrolysis using nuclear energy, but this method of production is also sometimes referred to as red or purple.

Turquois hydrogen is the name waiting for hydrogen to be produced by pyrolysis, which is a very high-temperature process that creates carbon in a granular form that is easy to store.

13.7 The Race to Deliver Net Zero

The race to deliver net zero in the CO_2 game has introduced hydrogen as a catalyst, or key factor, in being able to achieve the net zero challenge in an acceptable time frame. This chapter has looked at the properties of hydrogen, the way it is made, and the efforts that are being made to make the revolution happen. Clearly, it will only succeed if the energy to manufacture the hydrogen comes from renewable solar and wind or from new nuclear capacity about to come onstream. It can, and probably will, continue to be made using hydrocarbon feedstock, such as methane, but this must be accompanied by CCS for it to contribute to net zero.

The revolution needs to be massive to replace the enormous quantity of fossil fuel in current use. There appears to be an abundance of solar energy and wind available on this planet of ours, and the developments in the interior of Australia, Saudi Arabia, Morocco, and elsewhere give us hope that solar and wind will be up to the challenge. Solar panels produced 635,000 MW in 2019, about 4% of the world's energy production, and it is currently growing at over 20% per annum. If this rate continues for the next few decades, at a compound rate, then our energy problems will be solved. Nuclear power, with small modular reactors and the high temperatures that we can expect from Generation IV designs, should make a significant contribution to the battle to achieve net zero. Unfortunately, they seem to be very slow getting off the ground, with none of them yet operational.

Further Reading

The World Wide Web gives us rapid access to information in a flash. We can put any subject into a search engine, and it will lead us directly to many sources of information. To complete this section on further reading I have done just that, but I have filtered out and retained what I consider to be the best sources of information. There are several training documents, written to help students entering the nuclear industry, and I list these first of all. I then list some of the books that cover most of the reactor physics. Finally, I cover a list of topics for each chapter. I think Wikipedia need a special mention because it is usually at the top of the list as a competent source of information, on merit. Wikipedia also has numerous bibliographic references at the end of each article giving further depth to the search for knowledge. I suggest you use the search engine of your choice or the links (URLs) that offer an immediate face-to-face with the information you are looking for. Some of the links may require a ctrl+ click.

Training Documents

TRIGA Notes an IAEA training manual Microsoft Word—TRIGA_Physics_Kinetics_final_version (iaea.org)

US Office of Nuclear Energy. How Does a Nuclear Reactor Work, March 2021. https://www.energy.gov/ne/articles/nuclear-101-how-does-nuclear-reactor-work#:~:text=The%20water%20in%20the%20core,and%20the%20process%20is%20repeated

World Nuclear Association, How does a nuclear reactor work, https://www.world-nuclear.org/nuclear-essentials/how-does-a-nuclear-reactor-work.aspx

Books

J Baggot, *The Quantum Story*, Oxford University Press, 2011.

PM Bellan, *Fundamentals of Plasma Physics*, an e-book, https://library.uoh.edu.iq/admin/ebooks/80023-fundamentals-of-plasma-physics—paul-m.-bellan.pdf

SA Holgate, *Nuclear Fusion: The Race to Build A Mini-Sun On Earth*, Icon Books, 2022.

D Jakeman, *Physics of Nuclear Reactors*, Good, 1966.

T Kerlin and B Upadhyaya, *Dynamics and Control of Nuclear Reactors*, Academic Press, 2019.

JR Lamarsh, *Introduction to Nuclear Reactor Theory*, American Nuclear Society, 2002.

KO Ott and RJ Neuhold, *Introductory Nuclear Reactor Dynamics*, American Nuclear Society, 1985.

C Tucker, *How to Drive A Nuclear Reactor*, Springer, 2019.

Chapter 1

Capacity factors: US Energy Information Administration, EIA, for tables of capacity factors.

Chernobyl disaster: comprehensive account, Wikipedia https://en.wikipedia.org/wiki/Chernobyl_disaster

PWR, BWR CANDU and other reactor types: https://world-nuclear.org/information-library/nuclear-fuel-cycle/nuclear-power-reactors/nuclear-power-reactors.aspx

Fermi pile CP1: Chicago Pile-1, Wikipedia.

Manhattan project: Wikipedia.

Atoms for Peace: Wikipedia.

Fracking: Wikipedia.

Pebble-bed reactor: Wikipedia.

Euratom: https://eur-lex.europa.eu/EN/legal-content/summary/treaty-on-the-european-atomic-energy-community-euratom.html

Molten salt: Wikipedia.

Windscale fire: Wikipedia and 'a brief history of the Windscale Fire' on YouTube.

Chapter 2

Pauli exclusion principle: Wikipedia.

Nuclear forces: the Nobel Prize site has an excellent article on all the forces in nature: https://www.nobelprize.org/prizes/themes/forces/#:~:text=The%20Strong%20Nuclear%20Force%20is,radioactive%20decay%20of%20certain%20nuclei

Energy and mass units: A good source of information is the book referred to as Kaye and Laby, Tables of Physical and Chemical Constants. It has sections on units and fundamental constants.

Antimatter: Encyclopedia Brittanica, an easy-to-read overview, covering pair production and annihilation.

Radioactive decay: Wikipedia covers everything, including the history.

Spontaneous fission and other types of radioactive decay: Australian Gov. Arpansa, https://www.arpansa.gov.au/understanding-radiation/what-is-radiation/ionising-radiation/radiation-decay

Chapter 3

The uncertainty principle: The Guardian Newspaper, an article worth reading. https://www.theguardian.com/science/2013/nov/10/what-is-heisenbergs-uncertainty-principle

Determinism: Wikipedia, covers the philosophy of determinism.

Spin: Chapter 6 on 'The Self-Rotating Electron' in Jim Baggot's book, *The Quantum Story*.

Parity particle physics: Encyclopedia Brittanica, easy read that includes parity not being conserved, a Nobel Prize discovery by Lee and Yang.

Neutrino: all things neutrino, describes the neutrino family, https://neutrinos.fnal.gov/whats-a-neutrino/

Chapter 4

The discovery of the battery by Volta: an adequate description, https://ethw.org/Milestones:Volta%27s_Electrical_Battery_Invention,_1799#:~:text=In%201799%2C%20Alessandro%20Volta%20developed,immersed%20in%20a%20chemical%20solution

Maxwell's equations: a good explanation followed by mathematics, https://www.fiberoptics4sale.com/blogs/electromagnetic-optics/a-plain-explanation-of-maxwells-equations

Measuring the velocity of light: three methods by Foucault, Fizeau, and Michaelson.

A simple explanation of the Foucault method with diagrams, https://www.phys.ksu.edu/personal/rprice/SpeedofLight.pdf

Fizeau toothed wheel diagram, https://kids.britannica.com/students/assembly/view/73199

Michaelson Morley experiment: Wikipedia.

The Ether: a theoretical substance, https://www.britannica.com/science/ether-theoretical-substance

Standards of mass length and time: definitions of all the SI base units, https://physics.nist.gov/cuu/Units/current.html

The Kibble balance: Wikipedia.

Chapter 5

Nuclear liquid drop model: excellent and readable account, https://www.oxfordreference.com/display/10.1093/oi/authority.20110803100108421;jsessionid=4D06324989B695D21864E395DA997F9D

Discovery of fission: a relaxing read, https://www.aps.org/publications/apsnews/200712/physicshistory.cfm.

Lise Meitner, a full account of her life, https://www.atomicarchive.com/resources/biographies/meitner.html

Neils Bohr, https://www.atomicarchive.com/resources/biographies/bohr.html

Nuclear fission: Wikipedia, https://en.wikipedia.org/wiki/Nuclear_fission

Pair production: Wikipedia, https://en.wikipedia.org/wiki/Pair_production, best of many options.

Compton scattering: Wikipedia, https://en.wikipedia.org/wiki/Compton_scattering.

Delayed neutrons: Wikipedia, https://en.wikipedia.org/wiki/Delayed_neutron#:~:text=In%20nuclear%20engineering%2C%20a%20delayed,minutes%20after%20the%20fission%20event

The energy of fission: Wikipedia, https://en.wikipedia.org/wiki/Nuclear_fission. This article is comprehensive on many aspects of fission with keyword links to sub-topics.

Decay heat: See power reactors in shutdown in Wikipedia, https://en.wikipedia.org/wiki/Decay_heat#Power_reactors_in_shutdown

Nuclear chain reaction: Wikipedia, https://en.wikipedia.org/wiki/Nuclear_chain_reaction

Chapter 6

Fick's law and diffusion: Wikipedia, https://en.wikipedia.org/wiki/Fick%27s_laws_of_diffusion

https://www.nuclear-power.com/nuclear-power/reactor-physics/neutron-diffusion-theory/

Transport theory: Malaysia University article, http://www2.fizik.usm.my/mromar/trimon/chapter2.pdf

Reactivity: Nuclear Power for Everybody, https://www.nuclear-power.com/nuclear-power/reactor-physics/nuclear-fission-chain-reaction/reactivity/

Nuclear fuel: Energy Education, https://energyeducation.ca/encyclopedia/Nuclear_fuel#:~:text=Nuclear%20fuel%20is%20the%20fuel,%2D235%20and%20plutonium%2D239

Nuclear reactor moderators: What is a nuclear moderator?

By Dr Nick Touran, PhD, PE, 2007-04-21, Updated 16 December 2022, Reading time: 4 minutes, https://whatisnuclear.com/moderation.html

Nuclear reactor coolants: Nuclear Reactor Coolants, Suraya Omar, 14 February 2011, http://large.stanford.edu/courses/2011/ph241/omar1/

Nuclear reactor poisons: see neutron poisons, Wikipedia, https://en.wikipedia.org/wiki/Neutron_poisonXenon Poisoning: Xenon-135

Reactor poisoning: Stanford U, Khalid Alnoaimi, 15 March 2014, http://large.stanford.edu/courses/2014/ph241/alnoaimi2/

Magnox reactors: IAEA, A long and comprehensive account of these reactors, https://inis.iaea.org/collection/NCLCollectionStore/_Public/30/052/30052480.pdf

Fast reactors: A comprehensive up-to-date report on the forthcoming fast reactor by the World Nuclear Association: https://world-nuclear.org/information-library/current-and-future-generation/fast-neutron-reactors.aspx

Chapter 7

Oklo: A very readable story, https://www.iaea.org/newscenter/news/meet-oklo-the-earths-two-billion-year-old-only-known-natural-nuclear-reactor

Fermi's first reactor CP1:Wikipedia, Chicago Pile 1, https://en.wikipedia.org/wiki/Chicago_Pile-1

Reactor pressure vessel: Wikipedia, https://en.wikipedia.org/wiki/Reactor_pressure_vessel

The pressuriser: Wikipedia, pressuriser (nuclear power), https://en.wikipedia.org/wiki/Pressurizer_(nuclear_power)

The steam generator: Wikipedia, Steam generator (nuclear Power), https://en.wikipedia.org/wiki/Steam_generator_(nuclear_power)

The boron loop: See Colin Tucker, *How to Drive a Nuclear Reactor*, p. 122.

Neutron detectors: Wikipedia, https://en.wikipedia.org/wiki/Neutron_detection. See the references for each specific type of detector.

N16 Method: An IAEA report giving a comprehensive account of the N^{16} method for the determination of reactor power. Utilisation N^{16} in Nuclear Power Plants, Hhkan Mattsson, Farshid Owrang, Anders Nordlund, https://inis.iaea.org/collection/NCLCollectionStore/_Public/35/032/35032555.pdf

Fuel temperature coefficient FTC: Wikipedia, Fuel temperature coefficient of reactivity, https://en.wikipedia.org/wiki/Fuel_temperature_coefficient_of_reactivity#:~:text=Fuel%20temperature%20coefficient%20of%20reactivity%20is%20the%20change%20in%20reactivity,as%20the%20fuel%20temperature%20increases

Moderator temperature coefficient MTC: Nuclear power.com, https://www.nuclear-power.com/nuclear-power/reactor-physics/nuclear-fission-chain-reaction/reactivity-coefficients-reactivity-feedbacks/moderator-temperature-coefficient-mtc/

Void coefficient: Wikipedia, https://en.wikipedia.org/wiki/Void_coefficient#:~:text=In%20nuclear%20engineering%2C%20the%20void,the%20reactor%20moderator%20or%20coolant

Reactor start up procedure: TECHNICAL UNIVERSITY DRESDEN Institute of Power Engineering Training Reactor. A comprehensive account is given in this training manual, https://tu-dresden.de/ing/maschinenwesen/iet/wket/ressourcen/dateien/akr2/Lehrmaterialien/start_e.pdf?lang=en

Chapter 8

Chernobyl: Wikipedia, Chernobyl Disaster, https://en.wikipedia.org/wiki/Chernobyl_disaster

Windscale fire: Wikipedia, https://en.wikipedia.org/wiki/Windscale_fire

Brown's Ferry: Nuclear Engineering International, remembering the Browns Ferry incident, an interesting account. https://www.neimagazine.com/features/featureremembering-the-browns-ferry-fire-40-years-on-4578707/

Three Mile Island: Wikipedia, Three Mile Island Accident, https://en.wikipedia.org/wiki/Three_Mile_Island_accident

Gray and Sievert: US National Library of Medicine, well-explained, https://www.ncbi.nlm.nih.gov/pmc/articles/PMC9959072/#:~:text=The%20Sievert%20is%20the%20most,as%20in%20a%20radiation%20emergency

nuclear regulators: US NRC, https://www.nrc.gov/reactors/operating.html

Passive nuclear safety: Wikipedia, https://en.wikipedia.org/wiki/Passive_nuclear_safety#:~:text=Passive%20nuclear%20safety%20is%20a,particular%20type%20of%20emergency%20

Loss of coolant LOCA: Wikipedia, Loss-of-coolant accident, https://en.wikipedia.org/wiki/Loss-of-coolant_accident

Earthquake risk: World Nuclear Association, gives an explanation and detailed account of recorded earthquake incidents at nuclear plants, easy read. https://world-nuclear.org/information-library/safety-and-security/safety-of-plants/nuclear-power-plants-and-earthquakes.aspx

Tsunami risk: IAEA, Tsunami case studies, https://www.iaea.org/newscenter/news/risks-nuclear-reactors-scrutinized-tsunamis-wake

World Nuclear Association report on the Daiichi Incident, https://world-nuclear.org/information-library/safety-and-security/safety-of-plants/fukushima-daiichi-accident.aspx

Radiation health hazard: Health effects of radiation, with links to sub-topics, https://www.cdc.gov/nceh/radiation/health.html#:~:text=How%20Radiation%20Affects%20Your%20Body,to%20cancer%20later%20in%20life

Chapter 9

Nuclear fuel cycle: World Nuclear Association overview, https://world-nuclear.org/information-library/nuclear-fuel-cycle/introduction/nuclear-fuel-cycle-overview.aspx

Uranium hexafluoride (HEX): Wikipedia, https://en.wikipedia.org/wiki/Uranium_hexafluoride#Properties

Gas centrifuge: Wikipedia, https://en.wikipedia.org/wiki/Gas_centrifuge#Separative_work_units

Uranium enrichment: World Nuclear Association, https://world-nuclear.org/information-library/nuclear-fuel-cycle/conversion-enrichment-and-fabrication/uranium-enrichment.aspx

Fuel fabrication: World Nuclear Association, https://world-nuclear.org/information-library/nuclear-fuel-cycle/conversion-enrichment-and-fabrication/fuel-fabrication.aspx

spent fuel ponds: Wikipedia, https://en.wikipedia.org/wiki/Spent_fuel_pool

Cherenkov radiation: IAEA, What is Cherenkov radiation, https://www.iaea.org/newscenter/news/what-is-cherenkov-radiation#:~:text=Cherenkov%20radiation%20is%20a%20form,light%20in%20a%20specific%20medium

Nuclear reprocessing: Wikipedia, https://en.wikipedia.org/wiki/Nuclear_reprocessing

PUREX process: Wikipedia, https://en.wikipedia.org/wiki/PUREX

Mixer settlers: Wikipedia, https://en.wikipedia.org/wiki/Mixer-settler#:~:text=Mixer%20settlers%20are%20a%20class,phases%20to%20separate%20by%20gravity

Vitrification: Nuclear waste, a concise account includes vitrification, https://hps.org/publicinformation/ate/q10009.html#:~:text=Vitrification%20is%20a%20process%20used,%2C%20sand)%20and%20then%20calcined

Nuclear waste: World Nuclear Association, Radioactive Waste Management, https://world-nuclear.org/information-library/nuclear-fuel-cycle/nuclear-wastes/radioactive-waste-management.aspx#:~:text=Radioactive%20waste%20includes%20any%20material,plutonium%20%E2%80%93%20are%20categorized%20as%20waste

Chapter 10

Euratom: European Parliament, https://www.europarl.europa.eu/about-parliament/en/in-the-past/the-parliament-and-the-treaties/euratom-treaty

NPT: Wikipedia, Treaty on the Non-Proliferation of Nuclear Weapons, https://en.wikipedia.org/wiki/Treaty_on_the_Non-Proliferation_of_Nuclear_Weapons

IAEA: Wikipedia, https://en.wikipedia.org/wiki/International_Atomic_Energy_Agency

Nuclear Safeguards: UK Gov. A short article with excellent links in 'see also', https://www.onr.org.uk/safeguards/

UK Compliance: ONR Nuclear Material Accountancy, Control, and Safeguards Assessment Principles (ONMACS), https://www.onr.org.uk/operational/other/onr-cnss-man-001.pdf

Chapter 11

Generation IV reactors: World Nuclear Association, https://world-nuclear.org/information-library/nuclear-fuel-cycle/nuclear-power-reactors/generation-iv-nuclear-reactors.aspx

Small Modular Reactors SMR: IAEA, What are Small Modular Reactors, https://www.iaea.org/newscenter/news/what-are-small-modular-reactors-smrs

Breeder reactor: Wikipedia, https://en.wikipedia.org/wiki/Breeder_reactor

Fast reactors: IAEA, https://www.iaea.org/topics/fast-reactors

Molten salt reactors: World Nuclear Association, https://www.world-nuclear.org/information-library/current-and-future-generation/molten-salt-reactors.aspx

Burning nuclear waste: The waste-burning Stable Salt Reactor, NE International, https://www.neimagazine.com/features/featurethe-waste-burning-stable-salt-reactor-9563796/

Levelized cost of electricity LCOE: Wikipedia, https://en.wikipedia.org/wiki/Levelized_cost_of_
electricity#:~:text=The%20levelized%20cost%20of%20electricity,generation%20on%20a%20
consistent%20basis

Chapter 12

Nuclear fusion: Wikipedia, https://en.wikipedia.org/wiki/Nuclear_fusion#:~:text=Nuclear%20
fusion%20is%20a%20reaction,particles%20(neutrons%20or%20protons)

Tokamak: Wikipedia, https://en.wikipedia.org/wiki/Tokamak

Inertial confinement fusion: Wikipedia, https://en.wikipedia.org/wiki/Inertial_confinement_
fusion

Tritium breeding: use ctrl + to access sawan_abdou_2006_FED[8512].pdf an excellent account of
breeding and fusion problems in general.

Divertor: ITER description of yhe divertor with related articles, https://www.iter.org/mach/Divertor

STEP: STEP information hub, https://step.ukaea.uk/

Plasma: P Gibbon, An introduction to plasma physics, https://cds.cern.ch/record/2203630/files/
1418884_51-65.pdf

Spherical tokamaks: Wikipedia, https://en.wikipedia.org/wiki/Spherical_tokamak

Chapter 13

Hydrogen production: Wikipedia, https://en.wikipedia.org/wiki/Hydrogen_production

Electrolyser: International Energy Agency, excellent and comprehensive, https://www.iea.org/
reports/electrolysers

Methane steam reforming: US Gov. Excellent, also covers water-gas shift, https://www.energy.gov/
eere/fuelcells/hydrogen-production-natural-gas-reforming#:~:text=In%20steam%2Dmethane
%20reforming%2C%20methane,for%20the%20reaction%20to%20proceed

Methane pyrolysis: US Energy, Excellent, also covers hydrogen colours, https://www.energy.gov/
sites/default/files/2021-09/h2-shot-summit-panel2-methane-pyrolysis.pdf

Water splitting: US Energy, excellent with links on how to do it, https://www.energy.gov/eere/
fuelcells/hydrogen-production-thermochemical-water-splitting#:~:text=How%20Does%20It
%20Work%3F,and%20produces%20hydrogen%20and%20oxygen

Carbon capture: National Grid Hub, https://www.nationalgrid.com/stories/energy-explained/
what-is-ccs-how-does-it-work

International Energy Agency, https://www.iea.org/fuels-and-technologies/carbon-capture-
utilisation-and-storage

Fuel cells: US Energy, https://www.energy.gov/eere/fuelcells/fuel-cells#:~:text=Fuel%20cells%20
work%20like%20batteries,)%E2%80%94sandwiched%20around%20an%20electrolyte

US Energy, fuel sell basics explaining the various types of fuel cell, https://www.energy.gov/eere/
fuelcells/fuel-cell-basics

Energy storage: Wikipedia with links to the many ways to store energy, https://en.wikipedia.org/
wiki/Energy_storage#:~:text=Energy%20storage%20is%20the%20capture,called%20an%20
accumulator%20or%20battery

Index

For the benefit of digital users, indexed terms that span two pages (e.g., 52–53) may, on occasion, appear on only one of those pages.

AGR, 5, 61
alpha particle, 17
ammonia, 132
AMR, 113
amu (atomic mass unit), 14
Anderson, Carl, 31
Angstrom, Anders, 27–28
annihilation, 16–17
antimatter, 16
ash, 124
Aswan dam, 2
Atoms for Peace, 7, 50
Australia, 130, 137

back-end, 97–98
Balmer series, 27–28
barium, 41
barn, 19
basalt rock, 136
Becquerel, Henri, 20, 31
becquerel unit, 22
beryllium, 124
beta decay, 31
Bethe, Hanse, 28
Bhadla Solar Park, 120
blanket, 124–125
Bletchley Park, 10
Bohr, Niels, 27–28, 41, 42
Boltzmann's equation, 52
boric acid, 74–75
boron, 62–63
    acid, 63, 79
    carbide, 62–63
    loading loop, 74–75
bosons, 25
Boyle, Robert, 131
Brayton cycle, 112
breeding blanket, 125
Brown's Ferry, 85–86
burning waste, 118
BWR, 5

cadmium, 62–63
Calder Hall, 5–6

californium, 23
Campbell channel, 67
CANDU, 5, 7, 60
capacity factor, 3
capture reactions, 46–47
carbon
    Carbon Capture and Storage CCS, 133–134
    Carbon Capture and Utilisation CCU, 135
    credits, 121
    dioxide, 1
    offsets, 121
Carlsberg Foundation, 42
Cavendish, Henry, 131
Cavendish Nuclear, 113–114
centrifuge, 98
ceramic pebbles, 59–60
Chadwick, James, 31–32
chain reaction, 50
Chapelcross, 6
Cherenkov Radiation, 102
Chernobyl, 88–89
China, 116
climate change, 1
closed cycle, 97–98
CND Campaign for Nuclear
        Disarmament, 7–8
Cockcroft, Sir John, 84–85
cold leg, 71, 83
COLEX lithium enrichment, 129–130
Compton, Arthur, 47
    scattering, 47
contracts of difference, 121
Control
    rods, 79, 89
    room, 78–79
COP Conference of Parties, 1
Copenhagen, 41
Coulomb barrier, 12
CP1, 54, 70
Critical, 54
    mass, 58
    sub-critical, 54
    super-critical, 54
cross over leg, 71

cross-sections, 18
  macroscopic, 19
  microscopic, 19
  partial, 19
  total, 19
Culham CCFE, 125–126
Curie, Marie, 20–23, 102
cyclohexane, 132

DAC design acceptance confirmation,
  111–112
Daiichi, 84, 89
Daini, 89
Dalton, 14
Davisson, Clinton, 24
De Broglie, Louis, 24
  wavelength, 24
Debye, Peter, 28
decay
  chain, 22–23
  constant, 22
  heat, 82
delayed neutrons, 48–49, 56, 69–70
DEMO, 125
Determinism, 25–26
deuterium, 50
deuteron, 17, 122
diffusion theory, 52
Dirac, Paul, 16, 25–26, 31
Divertor, 124
doppler effect, 76
Dounreay, 61–62

earthquake, 81, 89
EAST, 125
Einstein's theory, 14, 36
elastic scattering, 46
electrolyser, 132
electrolysis, 132
electromagnetism, 36
electrometallurgy, 119
electron volt, 14
engineering materials, 124
enrichment of uranium, 98
erbium, 65
ether, 37
Euratom, 107
EUROFER97, 124–125
exclusion principle, 26–27
exit channels, 25–26

fast
  breeder reactor, 61–62
  fission factor, 55

  no-leakage probability, 55
  reactor, 66–67, 112–113
Fat Man, 10
Fermi, Enrico, 20, 25–26
  Pile (CP1), 54, 70, 71, 90–91
Fermions, 11
fertile, 45, 116
Feynman, Richard, 34
Fick's Law, 52
fissile, 45
fission
  decay heat, 50
  energy, 49–50
  fragments, 48
fissionable, 45
Fizeau, 36–37
FLiBe, 62, 116–117
Fort Belvoir, 6–7
fossil fuel, 2
Foucault, 36–37
fracking, 107
Frank, Ilya, 102
Frisch, Otto, 41
front-end, 97–98
fuel
  assembly, 99–100
  cells, 136
  fabrication, 99–100
  pins, 99–100
  shuffling, 59
  temperature coefficient, FTC, 76
Fukushima, 89

gadolinium, 65
gamma decay, 12–13
Gamow, George, 30, 41
Geiger
  counter, 22
  Hans, 32–33
Generation IV reactors, 111
Geothermal, 3
Gerlach, Walther, 29
GIF Generation IV International Forum, 6
GLEEP, 6
global warming, 1, 131
GNEP Global Nuclear Energy Partnership, 118
Goudsmit, Samuel, 29
graphite, 60–61
gray, (radiation dose unit), 95–96
GridReserve, 117
Groves, Leslie, 8–9

hafnium, 65
Hahn, Otto, 41

half-life, 22
Handford, 6
health hazard, 92
heavy water, 7, 51, 69–70
Heisenberg, Werner, 25–26, 28
HEU high enriched uranium, 108
HEX, 98
Higgs boson, 25–26
HLW high level waste, 105–106
hot leg, 71
HTGR high temperature gas reactor, 7
Hydrocracking, 132
Hydrogen, 8, 131
  brown, 137
  colours, 138
  engines, 136
hydrometallurgy, 119

IAEA The International Atomic Energy
  Agency, 108
Iceland Deep Drilling Project (IDDP), 3
IEA International Energy Agency, 2–3
importance factor, 48–49
India, 116
inelastic scattering, 46
INES, 87
integral reactor, 112
internal conversion, 26
iodine, 85
ITER, 124–125

JET Joint European Torus, 123

K capture, 26
Kariba dam, 2
Kibble balance, 40
Korean KSTAR, 125
Kronig, Ralph de Laer, 29

LCOE Levelised Cost of Electricity, 119–120
Lee, 30
leptons, 32
LEU low enriched uranium, 108
liquid drop model, 41
lithium, 128
  hydride, 131–132
  inelastic scattering, 128
Little Boy, 10
LLW low level waste, 105
LOCA loss of coolant accident, 83, 88, 90
Lorenz, 29
Los Alamos, 9–10, 93
lutecium, 65
Lyman series, 27–28

Mach, Ernst, 28–29
magic dust, 136
magnetic moment, 29
magnox, 65–66, 100
Manhattan Project, 8
mass excess, 17–18
MAST Mega Ampere Spherical
  Tokamak, 125–126
MASTU upgrade, 126
Maxwell, James Clerk, 36–37
  Distributions, 43–44
MDEP multinational design evaluation
  programme, 119
mean free path (mfp), 19
Meitner, Lise, 32–33, 41
Melville, Thomas, 27–28
methane pyrolysis, 134
methanol, 132
Michaelson, 36–37
Michaelson and Morley, 37
mining, 98
mixer-settler, 103
moderation, 58–59
moderator, 46
  temperature coefficient, 77
molten salt, 112–114, 116–117
  coolants, 62
Moltex, 117
  FlexReactor, 117
Monte Carlo, 52, 57–58, 128
MOX, 59–60, 105
muon, 32
Murchison Hydrogen Renewables, 137
MYRRHA LFR Project, 68

N16, 75
NAK, 61–62
NEOM green hydrogen project, 137
neutral particles, 123–124
neutrino, 31
  anti-neutrino, 32
  sterile, 32
neutron
  delayed neutron fraction, 48–49
  delayed neutron Importance Factor, 49
  effective delayed neutron fraction, 48–49
  interactions, 46
  multiplication, 127
  multiplication factor, 53
  precursors, 48–49
  thermal, 43–44
Newton's rings, 15–16
NPT non-proliferation of nuclear
  weapons, 108

NRC USA nuclear regulatory
    commission, 87–88
NRPB national radiological protection
    board, 95
nuclear
    Energy Agency, NEA, 87
    potential well, 12
    reactor generations, 5
    safeguards, 108
    strong nuclear force, 11–12, 122
    waste, 55, 113
    weak nuclear force, 32
nucleons, 11

Oak Ridge, 7
obligations, 110
Oklo, 69
open cycle, 97–98
Oppenheimer, Robert, 8–9

pair production, 16–17, 47
parity, 26–27, 29–30
Paschen series, 27–28
passive safety, 84
Pauli
    exclusion principle, 11, 28–29
    Wolfgang, 28–29, 31
Peierls, Rudolf, 43
PEM polymer electrolyte membrane, 133
PFR Prototype Fast Reactor,
    114–115
phase diagram, 98
photo electric effect, 47
Photo Voltaic PV, 2–3
photons, 14–15
pitchblende, 20
Planck, Max, 14–15
Planck's constant, 24, 40
plasma, 123
plutonium breeding, 114
poisons, 46–47, 62, 79
    burnable, 62, 65
    control, 62
    unavoidable, 62
PORV pilot-operated relief valve, 87
positrons, 16, 31
potassium, 94
power, 3
    measurement, 75
pressure vessel, 72
pressuriser, 73
protactinium, 65
PUREX Plutonium Uranium Extraction, 103,
    118

PWR Pressurised Water Reactor, 5, 61, 112
pyrometallurgy, 119

Q value, 17–18
quanta, 14–15
quantum
    electrodynamics, 34
    mechanics, 23
    numbers, 26–27
quarks, 11

radioactivity, 20
radium, 41
RCP reactor coolant pumps, 73
RCS reactor control system, 80
RCS reactor coolant system, 87
Reactivity, 56, 69
    Dollar, 56–57
    Nile, 56–57
reactor
    control system, 80
    kinetics, 52–53
regulators, 82
relativity, 36
reprocessing, 95, 102–103
resonance escape probability, 55
resonance region, 77
RHS residual heat removal system, 86
robotics, 125
Rolls Royce, 111–112
Rontgen, Wilhelm, 20
RPS reactor protection system, 82–83
Rutherford, Earnest, 20–22
Rydberg, Johannes, 27–28

safe-by-shape, 58
samarium poisoning, 64
scattering, 12
Schrodinger, Erwin, 25–26
sea level, 1–2
secondary containment, 83–84, 89, 91–92
SFM special fissile materials, 107
Shippingport, 5–7
SI safety injection system, 83
Siberia, 130
sievert, 95–96
simultaneity, 38
SMR small modular reactor, 113
SMR steam methane reforming, 133
SoDA statement of design
    acceptability, 111–112
SOEC solid oxide electrolyte cell, 133
solar cells
    farms, 120

solar cells (*Continued*)
  panels, 2–3
  PV, 111
Sommerfeld, Arnold, 28
specific activity, 22
spent fuel, 100–101
  ponds, 102
spherical tokamak, 125–126
spin, 26–27, 29
spontaneous fission, 23
Stark effect, 27–28
steam generator, 73
STEP, 126
Stern, Otto, 29
Strassman, Fritz, 41
Suiso Frontier, 137
superconducting magnets, 124–125
Super-X-Divertor, 125–126
SUR start up rate, 79

Tamm, Igor, 102
tau, 32
TBR tritium breeding ratio, 127
thermal
  fission factor, 55
  non-leakage probability, 55
  utilisation factor, 55
thorium
  breeding, 115
  emanations, 20–22
THORP thermal oxide reprocessing plant, 103
Three Gorges dam, 2
Three Mile Island TMI, 87
Tokamak, 123–125
transport theory, 52
TRISO, 59–60, 118
tritium, 17, 65, 94
Tritium Breeding Ratio, 127
triton, 17, 122
tsunami, 81, 89
tungsten, 124
tunnelling, 30

U-battery, 113–114
Uhlenbeck, George, 29
uncertainty principle, 24
UNDO Un-D0 CO2, 136
UNFCCC, 1
unified model, 26–27
uranium, 93–94
  hexafluoride, 98
  mill, 98
UREX, 119

VC void coefficient, 78
velocity of light, 36–37
vitrification, 105–106
Volta, 36

water-gas shift reaction, 133
water splitting, 135
Watt spectrum, 43–44
wave function, 26
wave mechanics, 23
Worley design, 137
Wigner energy, 85
wind turbines, 2, 111
Windscale, 6, 95
  fire, 7–8, 84–85

X10 reactor Oak Ridge, 6
Xenon
  poisoning, 63
  precluded start-up, 64–65
Xlinks Morocco-UK Power Project, 120
X rays, 12–13

Yang, 30
yellow cake, 98

Zeeman effect, 27–28
  anomalous, 28–29
ZIP, 70
zircaloy, 66
zirconium, 53